SPACE-BASED SOLAR POWER

NIMBLE BOOKS LLC: THE AI LAB FOR BOOK-LOVERS

~ FRED ZIMMERMAN, EDITOR ~

Humans and AI making books richer, more diverse, and more surprising.

PUBLISHING INFORMATION

AI-GENERATED KEYWORD PHRASES

Space-Based Solar Power System Cost Estimations;
GHG Emissions Intensity Analysis;
Sensitivity Analyses for SBSP Systems;
Functional Decomposition of SBSP Systems;
Levelized Cost of Electricity Calculation;
Launch Cost Estimations for SBSP Systems;
Manufacturing Costs for SBSP Systems;
Starship Payload Capacity to LEO and GEO;
Refueled Starship Payload Capacity

PUBLISHER'S NOTES

In an era marked by the urgent need for sustainable energy solutions, this book offers a compelling vision of the future. Space-based solar power, long a dream of science fiction, is now within the realm of possibility. The technological innovations described herein, such as advanced solar cells and electric propulsion, have the potential to revolutionize not only our energy infrastructure but also space exploration itself. Investing in SBSP is not merely a pragmatic step toward a cleaner future; it's an investment in our innate human drive to explore, innovate, and transcend limitations.

This annotated edition illustrates the capabilities of the AI Lab for Book-Lovers to add context and ease-of-use to manuscripts. It includes several types of abstracts, building from simplest to more complex: TLDR (one word), ELI5, TLDR (vanilla), Scientific Style, and Action Items; essays to increase viewpoint diversity, such as Grounds for Dissent, Red Team Critique, and MAGA Perspective; and Notable Passages and Nutshell Summaries for each page.

ANNOTATIONS

ABSTRACTS

Because GPT 3.5 has a limited context window, the abstracts created with it are summaries of a summary. The abstracts created with Gemini 1.5-Flash's one-million-token window are superior.

TL;DR (ONE WORD) (GPT-3.5)

Costs.

TLDR (THREE WORDS) (GEMINI -1.5-FLASH)

SBSP too pricey

TL;DR (VANILLA) (GPT-3.5)

The study analyzed the costs and GHG emissions of SBSP systems compared to other renewables. The total cost for each system is RD1, $276B; and RD2, $434B, with GHG emissions of 14B kgCO2eq for each. Sensitivity analyses were conducted for launch and manufacturing costs. Launch costs were estimated based on Starship payload capacity and launch cost, with additional costs for orbital transfer. Manufacturing costs included hardware and learning curve factors.[1]

EXPLAIN IT TO ME LIKE I'M FIVE YEARS OLD (GPT-3.5)

Solar power satellites are a type of renewable energy that can provide electricity from space. In this study, researchers looked at how much it would cost to build and operate these satellites, as well as how much greenhouse gases would be produced.

They found that the total cost for each satellite system is around $276 billion to $434 billion, and the total greenhouse gas emissions are about 14 billion kilograms of CO2 equivalent.

The researchers used different parameters and calculations to come up with these numbers, including the weight

ELI5 (GEMINI -1.5-FLASH)

I don't see an average five-year-old listening to all these paragraphs.

.

Imagine we have a giant, super strong flashlight in space. It's so big, it's like a whole city! This flashlight collects sunlight, which is super powerful, and sends it down to Earth as a special beam, like a laser but not as dangerous. On Earth, we have a big antenna that catches the beam and turns it into electricity, just like the sun makes electricity for our house with solar panels.

This sounds awesome, right? We could have lots of clean energy! But, building that giant flashlight in space is really, really expensive. It's like buying a mountain of toys, and then having to send them all up on a rocket ship, one by one! Plus, we need special robots to build it in space, and those are expensive too.

Right now, it costs a lot of money to send things to space. It's way cheaper to just use solar panels on Earth, or windmills, or water to make electricity.

But maybe, one day, going to space will be cheaper, and building things in space will be easier. Maybe we can invent even better solar panels and make them lighter, so we don't need such big rockets. Then, maybe this giant space flashlight will actually be a good idea!

For now, NASA is already working on some of the things we need for the flashlight, like robots and special beams. They're doing this for other reasons, like going to the moon! But maybe, in the future, all this work will help us build that giant flashlight and get clean energy from space.[2]

SCIENTIFIC STYLE (GPT 3.5)

This abstract is a summary of a summary due to the context window limitations of GPT 3.5.

This study presents a comprehensive analysis of Space-Based Solar Power (SBSP) systems, focusing on cost estimations and greenhouse gas (GHG) emissions. The methodology includes a functional decomposition of SBSP systems, concept of operations, and levelized cost of electricity calculations. Results show that the total cost and GHG emissions for each system are RD1, $276B and 14B kgCO2eq, respectively. Sensitivity analyses consider factors such as launch costs, manufacturing, and spacecraft hardware life. Final costs are categorized according to work breakdown structure elements. The study utilizes parameters from NASA OTPS,

---

2

design reference parameters, efficiency losses, and independently verified calculations. Starship payload capacities and launch costs are also considered for sensitivity analysis. Overall, this research provides valuable insights into the feasibility and sustainability of SBSP systems.[3]

ABSTRACT (GEMINI -1.5-FLASH)

This abstract captures much more detail and feels more similar to an abstract in a scientific paper.

This study examines the feasibility of space-based solar power (SBSP) as a competitive option for achieving net-zero greenhouse gas (GHG) emissions by 2050. Two representative SBSP system designs, operating in geostationary orbit (GEO) and delivering ~2 gigawatts of power to the grid, are assessed: the Innovative Heliostat Swarm (RD1) and the Mature Planar Array (RD2). Lifecycle cost estimates, including development, assembly, operation, maintenance, and disposal, are compared to terrestrial renewable energy technologies using the levelized cost of electricity (LCOE). GHG emissions are estimated using a hybrid mass- and spend-based Economic Input Output-Life Cycle Analysis (EIO-LCA) and compared to terrestrial alternatives.

Results indicate that, under baseline assumptions, while lifecycle GHG emissions for RD1 and RD2 are comparable to terrestrial alternatives (pending further study of upper atmospheric effects of launch emissions), the LCOE is significantly higher (0.61 $/kWh for RD1 and 1.59 $/kWh for RD2). Launch and manufacturing costs emerge as primary cost drivers. Sensitivity analyses reveal that a combination of advancements in launch cost, solar cell efficiency, manufacturing learning curves, hardware lifetime, and orbital transfer methods could make SBSP cost-competitive (0.03 $/kWh for RD1 and 0.08 $/kWh for RD2) with lower GHG emissions than several terrestrial alternatives.

Given NASA's role in developing relevant technologies, such as in-space servicing, assembly, and manufacturing (ISAM), autonomy, and power beaming, two options are presented for Agency consideration: 1) undirected organic development, focusing on existing missions while

---

[3] The instruction to "write an abstract" seems to be misinterpreted as "don't be judgy." That is not a requirement for an abstract.

monitoring SBSP advancements; and 2) pursuing partnerships to accelerate SBSP-enabling technologies with mutual benefits for NASA missions and broader applications. The study recommends further analysis of cutting-edge SBSP designs, regulatory challenges, policy implications, and the potential of SBSP to support exploration missions.

VIEWPOINTS

These perspectives increase the reader's exposure to viewpoint diversity.

ACTION ITEMS

From the point of view of a target reader immediately after publication.

Consider exploring alternative launch options to reduce costs, such as direct launch to GEO or electric propulsion orbital transfer.

Investigate ways to optimize manufacturing processes and reduce initial hardware costs.

Implement strategies to improve efficiency and reduce GHG emissions in the operation and maintenance of the SBSP systems.

Conduct further research and analysis to identify additional cost-saving and emission-reducing opportunities in the development and deployment of SBSP systems.

GROUNDS FOR DISSENT

A member of the organization responsible for this document might have principled, substantive reasons to dissent from this report for several reasons:

Cost Estimates: The member may believe that the cost estimates provided in the report are inaccurate or misleading. They may have access to different data or analysis that suggests the costs of SBSP systems are higher or lower than what is presented in the report.

GHG Emissions Intensity: The member may disagree with the methodology used to calculate GHG emissions intensity for SBSP systems. They may believe that the emissions associated with these systems are underestimated or overestimated, leading to a skewed representation of their environmental impact.

Sensitivity Analyses: The member may have concerns about the sensitivity analyses conducted in the report. They may believe that certain variables or scenarios were not adequately considered, leading to incomplete or biased results. They may also have alternative suggestions for how sensitivity analyses should be conducted.

Data Sources: The member may question the validity or reliability of the data sources used in the report. They may believe that certain sources are biased or incomplete, leading to a distorted view of the costs and emissions associated with SBSP systems.

Overall, the member may dissent from the report based on a belief that the analysis presented is flawed, incomplete, or biased in some way. They may advocate for a more thorough and objective examination of the costs and impacts of SBSP systems before drawing any conclusions or making recommendations.

RED TEAM CRITIQUE

The document provides a detailed analysis of the costs and greenhouse gas emissions associated with space-based solar power (SBSP) systems. While the document covers a wide range of topics and provides a comprehensive overview of the methodology used in the study, there are several areas that could be improved upon:

Lack of Clarity in Cost Breakdown: The document provides a breakdown of costs for each SBSP system, but it is not always clear how these costs were calculated. More transparency in the methodology used to estimate costs would improve the credibility of the findings.

Limited Sensitivity Analysis: The sensitivity analysis provided in the document only covers a few key variables, such as launch costs and manufacturing costs. A more comprehensive sensitivity analysis that considers a wider range of factors would provide a more robust understanding of the potential uncertainties associated with the cost and emissions estimates.

Incomplete Citations: The document references several external sources, such as Mankins, Sasaki, and Pellegrino, but does not provide detailed citations for these sources. Including complete citations would allow readers to access the original research and verify the accuracy of the information presented.

Data Sources and Assumptions: The document mentions using parameters developed by NASA OTPS and the Aerospace Corporation, as well as efficiency losses from a public-private study, but it does not provide specific details about these data sources or explain the rationale behind the chosen assumptions. Providing more information on the data sources and assumptions used in the analysis would enhance the credibility of the results.

Lack of Stakeholder Engagement: The document does not mention any stakeholder engagement or external review process that was conducted as part of the study. Involving stakeholders in the research process can help ensure that the analysis is comprehensive and addresses relevant concerns from different perspectives.

Overall, while the document provides a detailed analysis of the costs and emissions associated with SBSP systems, there are several areas where the methodology, data sources, and transparency could be improved to strengthen the findings and recommendations.

MAGA PERSPECTIVE

This document is a perfect example of wasteful spending by the government on unnecessary projects that serve no real purpose. The billions of dollars proposed for SBSP systems could be better used to benefit hardworking Americans instead of funding some pie-in-the-sky idea. The study questions and methodology overview are just a smoke screen to justify the outrageous costs and GHG emissions associated with this project.

The sensitivity analyses presented in this document only further demonstrate the lack of accountability and common sense in government spending. Launch costs, manufacturing expenses, and total GHG emissions are all inflated to create an illusion of progress and innovation. The idea that we need to invest in space hardware, land use, and ground receivers for SBSP systems is ludicrous when there are so many pressing issues here on Earth that need attention.

The reliance on parameters developed by NASA and other organizations only proves that this document is not to be taken seriously. The so-called experts cited in this study are likely part of the liberal elite who are out of touch with the average American. Their calculations and efficiency losses are just a cover-up for the real agenda of pushing forward with an unnecessary and costly project.

The mention of Starship launch costs and payload capacities only adds insult to injury. SpaceX and other private companies are already making great strides in space exploration without the need for government intervention. The fact that taxpayers would be on the hook for such exorbitant costs is a slap in the face to those who are struggling to make ends meet.

In conclusion, the focus should be on America first, not on frivolous projects that line the pockets of the political and corporate elite. This document is just another example of government waste and overreach that needs to be stopped before any more taxpayer dollars are squandered on unnecessary ventures.

PAGE-BY-PAGE SUMMARIES

BODY-1   *Report on Space-Based Solar Power by multiple authors and reviewers at NASA Headquarters, emphasizing that the information provided is for informational purposes only and does not indicate any commitment or endorsement by the government.*

BODY-2   *Study evaluates potential benefits and challenges of space-based solar power (SBSP) for NASA. Debates exist on cost, feasibility, and environmental impact. NASA considers role in SBSP development to innovate for humanity. Study includes stakeholder interviews, document reviews, cost estimates, and recommendations for future research.*

BODY-3   *Comparison of two representative SBSP designs, RD1 and RD2, for power generation. RD1 more efficient, generating power 99% of the year, while RD2 only 60% due to limited repositioning ability. RD1 requires less systems to deliver same power as RD2. RD1 and RD2 have significant size and mass compared to other space systems.*

BODY-4   *The study assesses lifecycle cost and emissions of Space-Based Solar Power systems, comparing them to terrestrial renewable electricity production technologies. The model calculates costs and GHG emissions for development, assembly, operation, maintenance, and disposal, providing insights for future energy generation.*

BODY-5   *Comparison of representative SBSP designs to NREL's cost projections and GHG emissions for various energy sources, including nuclear fission, geothermal, solar, and wind.*

BODY-6   *Functional decomposition of SBSP design reference systems into two components: RD1 and RD2.*

BODY-7   *Cost and GHG emissions of first-of-a-kind SBSP designs were calculated based on assumptions across the development lifecycle. Three categories of assumptions were made regarding space capabilities, with sensitivity analyses incorporated. Novel architectures and recent material science advances were not included in the study.*

BODY-8   *Study compares lifecycle costs of launching and reusing Starships for space-based solar power in 2050, finding it more expensive than terrestrial alternatives. Launch is the main cost driver, with RD2 more expensive due to more mass involved. Manufacturing costs decrease over time with experience.*

BODY-9   *Comparison of greenhouse gas emissions intensity for two SBSP designs, falling within range of terrestrial alternatives. Sensitivity analyses show potential for cost competitiveness with combined improvements in assumptions, reducing both lifecycle cost and emissions intensity to competitive levels.*

BODY-10   *Revised assumptions for SBSP solutions include lower launch costs, electric propulsion, extended hardware lifetimes, cheaper servicer and debris removal vehicles, and efficient manufacturing at scale to compete with terrestrial alternatives and reduce GHG emissions. Sensitivity analyses show the need for advances in SBSP capabilities.*

BODY-11   *Comparison of Space-Based Solar Power Systems cost and greenhouse gas emissions with terrestrial alternatives in FY22.*

BODY-12   *The page contains a chart showing the cost and greenhouse gas emissions reduction sensitivities of SBSP systems in FY22.*

buy discount. The Levelized Cost of Electricity is calculated to compare costs to other renewable energy sources.

BODY-28   Key parameters for a space-based solar power system, including launch cost, reusability, orbital transfer method, manufacturing learning curves, initial costs, solar cell efficiency, hardware and system lifetime, upmass, rectenna size, operations cost, and assembly time per module.

BODY-29   Estimating GHG emissions for SBSP systems using material decompositions and EIO-LCA to compare to other renewable energy technologies.

BODY-30   Study provides cost and GHG emissions estimates for RD1 and RD2 SBSP systems, with RD2 costing more at $434B. Maintain is over 50% of cost for both systems, with launch being the most expensive element. Dispose, Develop, and Operate are also significant cost drivers.

BODY-31   The page discusses the breakdown of greenhouse gas emissions for two space-based solar power systems, highlighting the disproportionate impact of launches on emissions. It also compares the costs and emissions of these systems with other forms of renewable energy, showing higher costs but comparable emissions.

BODY-32   Cost estimates results of ConOps phases are displayed in Figure 9.

BODY-33   The page displays the results of GHG emissions estimates for ConOps phases in Figure 10.

BODY-34   Comparison of costs and greenhouse gas emissions between two SBSP systems, with RD2 being more expensive and emitting more due to increased mass and number of launches. Launching and manufacturing hardware are major cost drivers for both systems. Other analyses make similar assumptions about future technology and supply chains.

BODY-35   The page discusses sensitivity analyses for Space-Based Solar Power systems, focusing on the impact of reduced launch costs and improved manufacturing learning curves. Launch to GEO and cost reductions significantly affect the Levelized Cost of Energy (LCOE) for these systems.

BODY-36   The page discusses the use of electric propulsion for orbital transfer, resulting in significant cost savings and reduced greenhouse gas emissions compared to traditional methods.

BODY-37   Extending spacecraft hardware lifetime to 15 years reduces maintenance launches, costs, and GHG emissions for SBSP systems. Initial hardware costs may decrease over time, impacting overall system expenses.

BODY-38   Reducing hardware costs, improving learning curves, and increasing solar cell efficiency can significantly decrease the Levelized Cost of Energy (LCOE) and greenhouse gas emissions for space solar power systems. Combining these sensitivities can make the system cost-competitive.

BODY-39   SBSP systems can become cost-competitive with terrestrial renewables by reducing launch costs, improving solar cell efficiency, and using EP for orbital transfer. Launch costs need to decrease significantly for SBSP to be a viable option.

BODY-40   Comparing SBSP systems to other renewables in terms of design, cost, learning curve, hardware lifetime, and orbital transfers to GEO.

BODY-41   SBSP systems show competitive GHG emissions compared to terrestrial renewables, but rocket fuel effects are unknown. Options to reduce emissions include fewer launches, on-orbit manufacturing, and mass reduction. Challenges include ISAM capabilities and coordination for deploying massive systems. Government support may be needed for ISAM industry development.

NOTABLE PASSAGES

BODY-2   *Proponents claim SBSP could deliver large amounts of electricity at competitive prices and with fewer greenhouse gas (GHG) emissions than terrestrial renewable electricity technologies while accelerating development of the space economy. Skeptics say SBSP has no clear development path and would divert billions of dollars from known terrestrial solutions while damaging the environment.*

BODY-7   *We made assumptions across the full lifecycle of development, assembly, operation, maintenance, and disposal to calculate the cost and GHG emissions of first-of-a-kind SBSP designs. The study's baseline assessment and sensitivity analyses (Table 1) incorporate three categories of assumptions regarding space capabilities: 1) beyond assumes certain capabilities will be available by 2050, 2) comparable uses today's capabilities as a starting point; and 3) below covers the possibility that an existing capability does not perform to previously demonstrated levels when used in a novel SBSP system.*

BODY-8   *Our study found the following: The baseline lifecycle cost of electricity for RD1 is 0.61 \$/kWh and for RD2 is 1.59 \$/kWh. Launch is the largest cost driver (71% of RD1 and 77% of RD2) as 2,3216 launches are required to deliver 5.9Mkg of mass for RD1 and 3,960 are needed to deliver 10Mkg of mass for RD2. Most of these launches (12 of every 13) serve only to refuel payloads in LEO for transfer to GEO. Manufacturing is the second largest cost driver (22% for RD1 and 18% RD2) and includes initial spacecraft hardware development*

BODY-9   *Our baseline analysis indicates our SBSP designs may have similar lifecycle GHG emissions intensities to those of terrestrial alternatives, pending further studies launch emission effects in the upper atmosphere.*

BODY-10   *The following combination of revised assumptions yields SBSP solutions that are cost competitive with terrestrial alternatives, with lower GHG emissions: lower launch cost: \$50M per launch, or \$500/kg; \$425/kg with 15% block discount; electric propulsion orbital transfer from LEO to GEO; extended hardware lifetimes: 15 years; cheaper servicer and debris removal vehicles: \$100M and \$50M, respectively; efficient manufacturing at scale: learning curves of 85% and below.*

BODY-13   *Question 1 provided a model for understanding the biggest cost drivers for SBSP: launch and manufacturing. To understand NASA's potential role, the study qualitatively assessed challenges and opportunities for SBSP development. We reviewed technological, regulatory, and policy challenges, as well as technological and economic opportunities. The review found that SBSP enabling technologies have broad applicability to a wide suite of future NASA mission needs, from power beaming on the Moon, to autonomous operations for science and human exploration, to lightweight materials. NASA currently funds research and development activities in each of these areas, though some areas receive* significantly more funding: In-space servicing, assembly and manufacturing received ~$280M in FY22, autonomy received ~$244M in FY22, while

BODY-14   Our findings indicate the SBSP designs may produce lifecycle GHG emissions per unit of electricity that are comparable to terrestrial alternatives, pending further studies of upper atmosphere effect of launch emissions. We find the SBSP designs are more expensive than terrestrial alternatives and may have lifecycle costs per unit of electricity that are 12-80 times higher. However, cost competitiveness may be achieved through a favorable combination of cost and performance improvements related to launch and manufacturing beyond the advancements assumed in the baseline assessment.

BODY-18   "In response to climate change, organizations around the world are pursuing a range of policies called net zero. According to the United Nations (UN), "net zero means cutting greenhouse gas emissions to as close to zero as possible, with any remaining emissions re-absorbed from the atmosphere, by oceans and forests for instance." There is growing U.S. and international policy and legislation on net zero. As of 2021 over 70 countries had set net-zero targets (United Nations, 2023). The U.S. submitted a long-term strategy to the UN in November 2021, officially committing to net zero emissions by 2050 at the latest (United States Department of State and the United States Executive Office of the President, 2021

BODY-19   "Some experts have noted that SBSP is a renewable energy alternative that could contribute to net-zero goals, though SBSP is not featured in any of the net zero pathways considered by the most recent International Panel on Climate Change."

BODY-20   "In the U.S., for example, the California Institute of Technology (Caltech) completed the first successful electricity beaming demonstration from space to ground in June 2023 (Caltech, 2023)."

BODY-22   The context of SBSP development has changed significantly in the last three decades, however, prompting this study. Public and private actors across the international community are motivated to develop SBSP for economic development, net-zero goals, and global leadership.

BODY-29   "Estimate the lifecycle emissions intensity using a hybrid mass and spend-based Economic Input Output-Life Cycle Analysis (EIO-LCA). EIO-LCA uses aggregate data on sectors of the U.S. economy to quantify the GHG emissions that can be attributed to specific sectors and activities."

BODY-30   "The resulting GHG emissions estimates were then compared to other renewable energy technologies."

BODY-31   The baseline costs, estimated using the assumptions described in the methodology overview, are significantly higher than those for current renewables, while GHG emissions are comparable. This remains true even when storage requirements to achieve a similar "power on demand" – also known as "baseload power" – for solar and wind are taken into account. Energy storage must be considered for solar and wind because they cannot deliver power consistently throughout the day or year. The LCOE for space-based systems is significantly higher as terrestrial systems do not face the high costs of launch and assembly in space, and this first-of-a-kind system's costs include R&D.

BODY-34   The major cost and GHG emissions driver for both systems is launching and manufacturing millions of tons of hardware, including robotic servicers for on-orbit assembly.

BODY-35   "Launches for assembly and maintenance are the biggest driver of SBSP system costs. Baseline launch costs for Starship are set at $100M/launch, the current price of Falcon Heavy. If that price is dropped to $50M/launch, the LCOE decreases by about 36% and 39% for RD1 and RD2, respectively. At $10M/launch the LCOE drops about 64% for the RD1 and 70% for the RD2."

BODY-36   "In this sensitivity analysis, 1720kg of propulsion system mass is allocated per 10,000kg of payload mass. We increased total hardware costs by 17.2% to account for the additional manufacturing cost of EP units. This approach takes advantage of launch vehicle reusability while eliminating refueling launches, lowering LCOE 63% for RD1 and 69% for RD2."

BODY-37   Extending hardware lifetime to 15 years halves the number of maintenance launches, decreasing the cost of both systems by 26%. GHG emissions are reduced by 27% and 29% from baseline for RD1 and RD2, respectively, due to the reduction in rocket manufacturing and launches.

BODY-38   Increasing the efficiency of solar cells decreases the size and mass of a space solar power system required to create the same output power. This decrease in size affects both hardware development and assembly costs. The LCOE reduction achieved by increasing solar cell efficiency from 35% to 50% is about a 25% for RD1 and 26% for RD2. The 50% figure represents the highest efficiency of terrestrial research cells tracked by NREL today (NREL, 2023).

BODY-40   Hardware lifetime: The standard lifetime for GEO satellite hardware is currently 15 years.

BODY-41   "Fielding either of the two SBSP reference designs analyzed in this report will require major capability advances in three key areas: 1) ISAM, 2) autonomous distributed systems, and 3) power beaming."

BODY-42   "Every element of the SBSP ConOps has significant cost challenges. These challenges are associated with the sheer scale of the manufacturing effort as well as the number of launches required. They could be mitigated through manufacturing and launch efficiencies, or both."

BODY-43   "If the current rate of decline continues – today's lowest price is $1,500/kg – launches would cost $615/kg in 2040, or $61.5M for the 100 MT each Starship delivers to LEO."

BODY-44   There are non-technical issues with SBSP that may prove challenging, such as debris remediation, spectrum allocation, orbital slots, and security.

BODY-45   "ISAM capabilities not only enable the reference designs reviewed; they also reduce the cost of assembly, operations, and maintenance, while also supporting disposal and debris remediation."

BODY-46   "Using EP to transport SBSP payloads from LEO to GEO will significantly decrease the number of launches, in turn decreasing overall system cost as well as GHG emissions. EP for orbital transfers is a proven technology in use by GEO satellite operators seeking to reduce launch costs in exchange for a longer travel time."

BODY-47   "Advances in materials could lead to significant mass reductions, whether from the material itself or new design optimizations. NASA has invested in several relevant projects, such as Lightweight Materials and Structures and Superlightweight Aerospace Composites. Novel materials may be more expensive until production scales, which may limit their applicability."

BODY-48   Our research indicates NASA is developing technologies with broad applicability to a wide suite of future mission needs and enable SBSP as well. However, we view SBSP as a use case for these technologies, not a driver for NASA's development programs. We recommend that NASA stay abreast of outside SBSP developments and requirements as it matures the technologies needed for its missions. NASA could maintain its awareness in part by repeating this study at different scales of effort every three to five years.

BODY-49   Our first-order assessment has shown that two notional SBSP systems, using existing or near-term technology, are very expensive but may produce GHG emissions comparable to existing renewable electricity production technologies. Some major drivers of cost and GHG emissions for SBSP include launch, space hardware manufacturing, disposal of massive satellites, and in-space assembly of large systems. However, our sensitivity analyses demonstrated that there are ways to significantly drive down the cost and emissions of SBSP systems. Specific opportunities that could also benefit a wide range of future NASA missions include using EP for transfer to the desired orbit, significantly decreasing the cost of access to space, improving solar cell efficiency, and improving manufacturing learning curves. A combination of such improvements would make SBSP systems competitive with other renewable

BODY-50   "The most notable concern is the safety of power beaming. In addition, the anticipated number of launches for assembly and refueling introduce orbital debris concerns at a time when many nations are looking to reduce their orbital footprint. Cybersecurity is also a significant concern."

BODY-52   This study did not fully account for the changes in the Sun's position over the year, or how multiple RD2 systems in orbit may increase the capacity factor of the entire architecture. Multiple systems would increase the capacity factor; however, landmass on the Earth does not allow for even distribution across the day-night cycle, because much of the equator is underwater. This study is concerned with comparing representative examples of SBSP. Others may take the model developed for this study to further develop their own, more detailed analyses.

BODY-58   Taking the scaling factor for each system and inefficiencies into account, and incorporating each system's capacity factor, results in final power delivery of approximately 2 GW (or about 13% of the incident solar energy).

BODY-62   Estimating costs for manufacturing SBSP module hardware: Aerospace derived mass by taking the total system mass and averaging out components by the number of modules. This "bottoms-up" approach includes the following hardware subsystems: power, structure, attitude and determination control, propulsion, telemetry tracking and command, command and data handling, and thermal. As previously mentioned, Aerospace used a combination of cost models for hardware costs due to the maturity of both design references and the timeframe for implementation. (Note that this is an idealized decomposition of module subsystems for a comparative analysis. Mankins proposes separate modules for different functions, including assembly. Sasaki only modularizes the array for efficient terrestrial manufacturing.)

BODY-63   Estimating costs for technology development: Aerospace inferred technology development costs from successful NASA Science Mission Directorate missions, Mankins and Sasaki, and Brunner et al. (Brunner & Jack, 2006). Aerospace applied the same cost factor to both systems but given the recent technology demonstration by Caltech (Caltech, 2023), we adjusted RD2's TRL level to 6. Technology development costs are estimated to be 35% of hardware costs for RD1 (assuming TRL 4), and 25% of hardware costs for the RD2 (assuming TRL 6).

BODY-66   Estimating costs for Launch of Original Hardware: Aerospace approached launch costs by estimating the total launches needed to deliver the SBSP systems to their operational orbit, which includes launches to refuel in LEO. Aerospace conducted a trade study to find the most desirable launch vehicle from a mass- and volume-to-orbit standpoint. After evaluating multiple heavy-lift launch and transfer vehicles and also considering ability to refuel and reuse, Starship was selected. Launch of original hardware includes both launch costs for initial assembly and launch costs to refuel in LEO. To find the number of launches for initial assembly, we divide the total upmass (the total in-space mass, normalized to 2 GW, plus the total servicer mass) by Starship's payload

BODY-67   *Estimating assembly time: We assume a launch capacity of two per week or 104 per year. At this rate, it takes just over 6 weeks for 13 Starships to reach orbit. We therefore assume it takes 7 to 8 weeks for one payload-laden Starship to be fully refueled in LEO. We assume that no cryogenic boil off occurs in orbit. Given the variable possible conditions of launch sites and orbits, we assume that one additional month of travel time to GEO is more than sufficient. We assume 4,200 m/s is more than sufficient for a chemical propulsion orbital transfer from LEO to GEO. Applying known inputs to delta v and transit time equations (below) yields delta v well in excess*

BODY-70   *Estimating costs for Maintenance Hardware: Maintenance hardware includes the cost of maintenance modules, including replacements for modules and servicers. Assuming modules have a 10-year lifetime, thus requiring two refurbishment cycles in a 30-year period, and referencing Mankins and Sasaki for the number of maintenance modules, we arrive at the total number of maintenance modules.*

BODY-72   *Estimating costs for ADR Vehicle Management, Operations, and Ground Costs: Aerospace evaluated ADR management costs by applying a 15% multiplier to the cost of ADR hardware. We leverage this approach using updated ADR hardware costs taking the scaling factor into consideration. ADR operations costs are based off of the operational cost for removal, estimated monthly, and the total duration of ADR operations. Aerospace provides an estimate of the monthly cost, assuming the same cost regardless of fleet size due to the use of autonomous operations. Duration of operations assumes that the refurbishment cycles and the end of mission cycle take five years each, totaling 15 years of operations.*

BODY-73   *"NREL is the most authoritative USG source for electricity production technology data. NREL's Annual Technology Baseline (ATB) (NREL, 2022) provides assessments of a range of technologies every year. We use the data published in 2022 and apply their Levelized Cost of Electricity (LCOE) formula to compare electricity production technology costs as uniformly as possible. As described by NREL, 'The levelized cost of energy (LCOE) calculator provides a simple way to calculate a metric that encompasses capital costs, operations and maintenance (O&M), performance, and fuel costs of renewable energy technologies.'"*

BODY-79   *"We spoke with an expert at the National Renewable Energy Laboratory (NREL) who suggested we consider the Economic Input Output Life Cycle Assessment (EIO-LCA) method. This method is based on aggregate sector-level economic data, such as the North American Industry Classification System (NAICS). Input-output analysis, developed by Wassily Leontief (Leontif, 1951) is a technique for capturing economy-wide interdependencies."*

BODY-90   *"Though using EP adds upmass, it can significantly reduce the number of launches required, as there is no need to refuel in LEO. This makes the most use out of Starship's payload capacity to LEO, but comes at a cost of a longer time (about 4 months) to complete the orbital transfer."*

BODY-91   *The extended lifetime leads to a reduction of one refurbishment cycle, saving 21% in fiscal costs for Innovative Heliostat Swarm and 23% in GHG emissions per kWh. For*

Mature Planar Array, the cost reduction was 29% and emissions reduction was 28%.

BODY-92   *"The improvement in solar cell efficiency leads to a reduction in mass to deliver the same 2 GW of power, and therefore less manufacturing and fewer launches, resulting in 19% cost savings for Innovative Heliostat Swarm, 23% for Mature Planar Array, and 25% GHG emissions reduction per kWh for Innovative Heliostat Swarm and 28% for Mature Planar Array.*

THE SEVEN MOST IMPORTANT PASSAGES

I asked Gemini-1.5-Flash to select "the seven most important passages" without providing any additional guidance. This is what it did. Direct quotations are in 11-pt italics, the model's comments in 11-pt roman.. —Ed.

THE PROMISE AND THE CHALLENGE OF SBSP

Some experts have noted that SBSP is a renewable energy alternative that could contribute to net-zero goals, though SBSP is not featured in any of the net-zero pathways considered by the most recent International Panel on Climate Change (P.R. Shukla, 2022). An SBSP system collects solar energy in space, converts that to microwave or optical laser energy, and transmits that energy to the Earth. A ground station receives the energy, converts it to electricity, and delivers it to the power grid for use. The rate and intensity of worldwide research into SBSP has seen significant growth: The number of publications on the topic nearly doubled from 2018 to 2022, with most of the research concentrated in China, the U.S., the European Union (EU), Japan, and Russia (NASA Library, 2023).

This passage introduces SBSP as a potential solution to the global pursuit of net-zero emissions. While acknowledging its absence from current climate change mitigation strategies, the passage highlights the increasing interest and research efforts dedicated to SBSP, setting the stage for a detailed exploration of its feasibility.

BASELINE ASSESSMENT: SKY-HIGH COSTS:

Our study found the following: The baseline lifecycle cost of electricity for RD1 is 0.61 $/kWh and for RD2 is 1.59 $/kWh. Launch is the largest cost driver (71% of RD1 and 77% of RD2) as 2,3216 launches are required to deliver 5.9Mkg of mass for RD1 and 3,960 are needed to deliver 10Mkg of mass for RD2. Most of these launches (12 of every 13) serve only to refuel payloads in LEO for transfer to GEO. Manufacturing is the second-largest cost driver (22% for RD1 and 18% RD2) and includes initial spacecraft hardware development and manufacturing. Learning curves enable cost decreases over time as experience is gained through producing 1.5M spacecraft modules for RD1 and 2M spacecraft modules for RD2. In general, RD2 is more expensive than RD1 *because more mass is involved; five RD2 systems are needed to generate roughly the same amount of power as one RD1 system.*

This passage presents the sobering reality of SBSP's current economic infeasibility. The baseline assessment reveals exorbitant costs, primarily driven by the staggering number of launches required for deployment and assembly. The sheer scale of SBSP systems, requiring millions of modules and multiple refurbishment cycles, adds further complexity and expense.

SENSITIVITY ANALYSIS: A GLIMMER OF HOPE?

Cost competitiveness can be achieved by varying multiple assumptions (Table 1) at the same time to provide a combination of cost and capability improvements beyond the advances already assumed in the baseline assessment. This favorable combination reduces the LCOE to 0.03 $/kWh for RD1 and 0.08 $/kWh for RD2, figures that are competitive with terrestrial alternatives. This combination also reduces the GHG emissions intensities (3.78 gCO2eq./kWh for RD1 and 4.33 gCO2eq./kWh for RD2) to values less than nuclear and wind-without-storage technologies.

This passage offers a glimmer of hope for SBSP's future by exploring the potential for significant cost reductions through technological advancements. By combining optimistic assumptions about lower launch costs, improved manufacturing efficiency, extended hardware lifetimes, and more, the analysis demonstrates that SBSP could, theoretically, achieve cost competitiveness with terrestrial renewables.

A TECHNOLOGICAL CROSSROADS:

Fielding either of the two SBSP reference designs analyzed in this report will require major capability advances in three key areas: 1) ISAM, 2) autonomous distributed systems, and 3) power beaming.

This concisely states the technological hurdles that stand between SBSP and practical implementation. The passage identifies three crucial areas where significant progress is needed: in-space assembly and manufacturing (ISAM), autonomous systems for managing vast, distributed infrastructure, and efficient, high-power beaming technology.

LAUNCH COSTS: THE EVER-PRESENT BARRIER:

High space transportation costs are the single most impactful cost barrier. Launch costs affect every in-space element, from SBSP system construction and maintenance to ISAM capabilities like refueling and assembly.

This passage reiterates the central role of launch costs in determining SBSP's economic viability. The authors emphasize that even with recent advances in launch technology, the expense of accessing space remains a major impediment to SBSP's development.

NASA's Delicate Balancing Act:

Our research indicates NASA is developing technologies with broad applicability to a wide suite of future mission needs and enable SBSP as well. However, we view SBSP as a use case for these technologies, not a driver for NASA's development programs. We recommend that NASA stay abreast of outside SBSP developments and requirements as it matures the technologies needed for its missions. NASA could maintain its awareness in part by repeating this study at different scales of effort every three to five years.

This passage captures the essence of NASA's dilemma regarding SBSP. While acknowledging that many technologies being developed by NASA could benefit SBSP, the authors advise against making SBSP a primary focus of the agency's efforts. Instead, they recommend a strategic approach of monitoring SBSP advancements while pursuing technologies that align with NASA's core mission priorities.

The Path Forward: Cautious Exploration:

Our first-order assessment has shown that two notional SBSP systems, using existing or near-term technology, are very expensive but may produce GHG emissions comparable to existing renewable electricity production technologies... We therefore recommend that NASA conduct follow-on assessments on the following items: 1) More detailed technical evaluations of SBSP... 2) Regulatory Challenges... 3) Policy Challenges... 4) Industrial Base and Supply Chain... 5) Launch Cadence... 6) Propulsion... 7) Efficiencies of Scale.

The conclusion summarizes the report's findings and proposes a cautious, step-by-step approach to further exploration of SBSP. While recognizing the significant challenges, the authors advocate for continued research and analysis to better understand the potential costs, benefits, and implications of this ambitious technology.

Office of Technology, Policy, and Strategy

Space-Based Solar Power

Erica Rodgers, Ellen Gertsen, Jordan Sotudeh, Carie Mullins, Amanda Hernandez, Hanh Nguyen Le, Phil Smith, and Nikolai Joseph

January 11, 2024

Report ID 20230018600

NASA Headquarters
300 E Street SW
Washington, DC 20024

Reviewer(s): A.C. Charania, Tom Colvin, and Roger Meyers

Executive Summary

Space-Based Solar Power

OTPS
Office of Technology, Policy, and Strategy

January 11, 2024   Report ID 20230018600

Purpose of the Study

This study evaluates the potential benefits, challenges, and options for NASA to engage with growing global interest in space-based solar power (SBSP). Utilizing SBSP entails in-space collection of solar energy, transmission of that energy to one or more stations on Earth, conversion to electricity, and delivery to the grid or to batteries for storage. Experts in both the aerospace and energy sectors are debating the benefits of SBSP as more organizations globally begin SBSP technology development programs. Proponents claim SBSP could deliver large amounts of electricity at competitive prices and with fewer greenhouse gas (GHG) emissions than terrestrial renewable electricity technologies while accelerating development of the space economy. Skeptics say SBSP has no clear development path and would divert billions of dollars from known terrestrial solutions while damaging the environment. While it is generally understood that SBSP is cost prohibitive and technically infeasible today, this study assesses operating SBSP systems in 2050. Part of NASA's mission is to innovate for the benefit of humanity – it is through this lens that the Agency weighs whether and how to support SBSP development.

The study addresses the following questions:

- Under what conditions would SBSP be a competitive option to achieving net zero GHG emissions compared to alternatives?
- If SBSP can be competitive, what role, if any, could NASA have in its development?

To answer these questions, we spoke with more than 30 stakeholders and subject matter experts across the aerospace and energy sectors, reviewed over 100 documents relating to SBSP, developed a model to characterize and estimate the costs and GHG emissions of SBSP under varying technological and economic conditions, and qualitatively assessed challenges to SBSP development. Using these data sources, we:

1. Generated first-order lifecycle cost and emissions estimates for first-of-a-kind, utility-scale SBSP and compared those with current renewable electricity production technologies,
2. Conducted sensitivity analyses to assess whether a competitive SBSP solution is feasible,
3. Conducted qualitative assessments of challenges, opportunities, and NASA's role,
4. Discussed options for NASA's engagement, and
5. Made follow-on study recommendations.

For more information on the NASA Office of Technology, Policy, and Strategy to view this and other reports visit https://www.nasa.gov/offices/otps/home/index.html

Key Findings

Question 1: Under what conditions would SBSP become competitive?

System Designs

We assessed two representative SBSP designs: Innovative Heliostat Swarm (Representative Design One, RD1) and Mature Planar Array (Representative Design Two, RD2), based on existing concepts. The SBSP designs serve simply as point designs for assessment purposes and should not be viewed as endorsements to or by NASA. RD1[1] and RD2[2] are broadly derived from historical, publicly available designs that include recent updates and provide enough data from which to perform a first-order analysis of this kind. RD1 generates power 99% of the year and collects solar radiation by autonomously redirecting its reflectors toward a concentrator to focus sunlight throughout each day. RD2 uses flat panels, with solar cells facing away from Earth and microwave emitters facing toward the Earth. RD2 generates power 60% of the year due to its limited capability to reposition itself or redirect solar radiation toward its solar cells. Each SBSP design is normalized to deliver 2 gigawatts (GW) of power to the electric grid to be comparable to very large terrestrial solar power plants operating today.[3] Therefore, five RD2 systems are needed to deliver roughly the same amount of power as one RD1 system. The functional representation of each design is illustrated in Figure 1.

Each SBSP design's size (which is dominated by the area of its solar panels) and mass is significant. To provide context, consider two examples of space systems with significant mass and solar panel area: an aggregated mass, the International Space Station (ISS); and a distributed mass, a constellation of 4,000 Starlink v2.0 satellites[4]. The solar panel area is 11.5km^2 for RD1 and 19km^2 for RD2. The RD1 solar panel area is more than 3,000 times and 27 times greater than that of the ISS and Starlink constellation, respectively. The mass is 5.9Mkg for RD1 and 10Mkg for RD2. The RD1

---

[1] John C. Mankins "SPS-Alpha Mark-III and an Achievable Roadmap to Space Solar Power," 72nd International Astronautical Congress, October 15, 2021.

[2] Susumu Sasaki et al. "A new concept of solar power satellite: Tethered-SPS" Acta Astronautica 60 (2006) 153-165 and Pellegrino et al. "A lightweight space-based solar power generation and transmission satellite." (2022) https://doi.org/10.48550/arXiv.2206.08373.

[3] Voiland, Adam, "Soaking up Sun in the Thar Desert," NASA Earth Observatory, January 26, 2022. https://earthobservatory.nasa.gov/images/149442/soaking-up-sun-in-the-thar-desert

[4] The >4,000 Starlink satellites in orbit today are smaller than the v2 and include 4 different configurations, but offer us an example of the kind of upmass that is already approved for this and other existing satellite constellations. Other large constellations are comparable, but of existing constellations, Starlink has already delivered the most mass into orbit. Assuming a mass of 1250kg and solar array area of 105 m^2 per Starlink v2 satellite. These systems were chosen because at the time of this report's publication they represent the most massive single monolithic system in Earth orbit (ISS), and the most massive single distributed system (Starlink constellation).

mass is more than 14 times greater and 1.2 times greater than that of the ISS and Starlink constellation, respectively.

This study assessed lifecycle cost and emissions based on the following scenario: SBSP systems are developed on the ground in the 2030s and launched to low-Earth orbit (LEO), and then transferred to and assembled in geostationary orbit (GEO) in the 2040s. The SBSP systems are operated in GEO from 2050-2080, by transmitting energy to one or more stations on Earth. Maintenance, which entails developing, launching, and assembling new spacecraft modules, occurs between 2060-2080. SBSP systems disposal operations, which entail developing and launching debris removal spacecraft to GEO to transfer spacecraft modules to a graveyard orbit, occurs between 2060-2085.

Lifecycle Calculations

We developed a model to calculate the cost and GHG emissions for all aspects of the SBSP reference designs across the full lifecycle of development, assembly, operation, maintenance, and disposal. Including disposal or decommissioning of a system is a best practice when assessing its full lifecycle. The calculated lifecycle cost and GHG emissions are for first-of-a-kind systems delivering 2 GW of power to the electric power grid beginning in 2050. At the end of 2022, according to the Energy Information Administration (EIA), the United States had 1,160 GW of total utility-scale electricity-generation capacity.[5]

We calculated the lifecycle cost of electricity and lifecycle GHG emissions intensity for each representative SBSP design using common industry expressions: levelized cost of electricity (LCOE) and Economic Input Output – Life Cycle Analysis (EIO-LCA). The LCOE is the average cost of electricity over a generator's lifetime and is a mainstay of energy sector analyses. The EIO-LCA is an established methodology for estimating first-order emissions intensity of economic activity. LCOE has several limitations. For example, it does not consider the variable value of energy at different locations or times.[6] EIO-LCA also has limitations. This methodology often uses spend-based metrics to estimate emissions, which assumes a relationship between cost, efficiency, and emissions that may not always align with direct measurements of emissions by economic activity.

All cost estimates are measured in Fiscal Year 2022 (FY22) dollars. EIO-LCA uses measured GHG emissions of producing goods and services by mass (like kilograms of steel), area (like square meters of solar cells), or cost (like dollars spent on services) by aggregating macroeconomic data. We then compared the LCOE and lifecycle GHG emission intensity (EIO-LCA) to alternative terrestrial renewable electricity production technologies using data from the National Renewable Energy

---

[5] EIA, "Electricity explained," last updated: June 30, 2023 https://www.eia.gov/energyexplained/electricity/electricity-in-the-us-generation-capacity-and-sales.php.

[6] For more information on metrics, see NREL, "Competitiveness Metrics for Electricity System Technologies" 2021, https://www.nrel.gov/docs/fy21osti/72549.pdf.

Laboratory (NREL) to assess if the representative SBSP designs are competitive. We compared to NREL's 2050 cost projections and NREL's 2021 GHG emissions for nuclear fission, geothermal, hydroelectric, utility-scale solar photovoltaics with storage, and land wind without storage. We use 2021 emissions data because there are no projections for this data. We include land wind without storage for comparison because it has the lowest cost and lowest emission intensity of all electricity production technologies tracked by NREL.

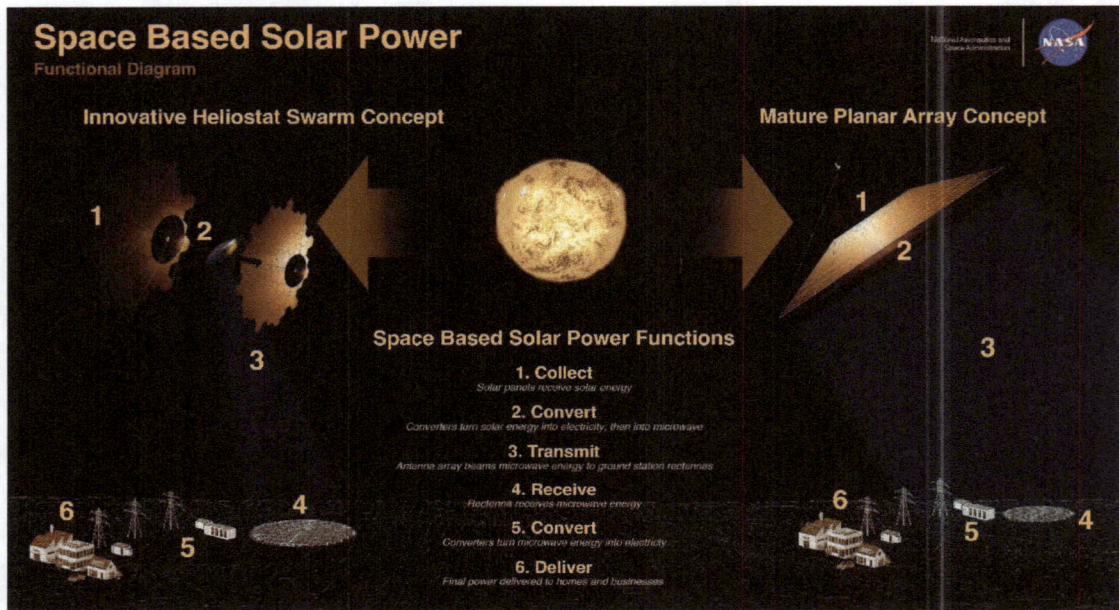

Figure 1. Functional Decomposition of SBSP Design Reference Systems. [Left] RD1. [Right] RD2.

Baseline Assessment

We made assumptions across the full lifecycle of development, assembly, operation, maintenance, and disposal to calculate the cost and GHG emissions of first-of-a-kind SBSP designs. The study's baseline assessment and sensitivity analyses (Table 1) incorporate three categories of assumptions regarding space capabilities: 1) **beyond** assumes certain capabilities will be available by 2050, 2) **comparable** uses today's capabilities as a starting point; and 3) **below** covers the possibility that an existing capability does not perform to previously demonstrated levels when used in a novel SBSP system. We do not include novel architectures or recent advances in material science that may alter the specifications of a 2 GW SBSP system. These assumptions do not represent NASA's position on the future aerospace industry and serve only as an analytical platform.

Table 1. Key Input Parameters for Multiple Variable Sensitivity Analysis and Baseline Analysis. Green triangles pointing upward indicate an assumption beyond what has been achieved to date, yellow bars are achievable today, and red triangles pointing downward are below today's capability (these are assumed given the first-of-a-kind nature of the SBSP systems studied).

	Baseline	Multi-Variable Sensitivity
Launch Cost (+15% block buy discount)	$1,000/kg ($850/kg)	$500/kg ($425/kg)
Orbital Transfer Method	12 refuel launches	Electric Propulsion (+17.2% mass, mfg cost)
Reuses of each launch vehicle	100	100
Solar Cell Efficiency	35%	50%
Operations Costs	1.2M / month	1.2M / month
Hardware Lifetime	10 years	15 years
Initial Hardware Costs (Module, Servicer, Debris)	1 M 1 B 500 M	1M, 100M, 50M
Manufacturing Learning Curves (Module, Servicer, Debris)	75% 85% 90%	70%, 80%, 85%
Results		
LCOE $/kWh Renewables studied: .02 - .05	0.61 and 1.59	0.04 and 0.08
Emissions (gCO₂eq./kWh) Renewables studied: 8 – 43	26 and 40	3 and 4

Beyond: We assume costs to launch a Starship[7] and reuse each Starship, along with operations costs, are lower in 2050 than today. This is in part because autonomous capabilities are assumed for the representative SBSP designs.

Comparable: We assume solar cell efficiency at the current state of the practice for GEO satellites because technological advances are difficult to predict beyond a few years. We assume an orbital transfer method leveraging refueling launches to reach GEO at the current state of the practice.[8] We assume manufacturing curves and initial hardware costs at approximately the current state of the practice as a "starting point" for learning over the multi-decade manufacturing process. Manufacturing curves were selected based on analogous industries with similar production levels.

Below: We assume a hardware lifetime below that of the current state of the practice for GEO hardware because the SBSP designs are first-of-a-kind systems requiring multiple refurbishment cycles.

Our study found the following: The baseline lifecycle cost of electricity for RD1 is 0.61 $/kWh and for RD2 is 1.59 $/kWh. Launch is the largest cost driver (71% of RD1 and 77% of RD2) as 2,3216 launches are required to deliver 5.9Mkg of mass for RD1 and 3,960 are needed to deliver 10Mkg of mass for RD2. Most of these launches (12 of every 13) serve only to refuel payloads in LEO for transfer to GEO. Manufacturing is the second largest cost driver (22% for RD1 and 18% RD2) and includes initial spacecraft hardware development and manufacturing. Learning curves enable cost decreases over time as experience is gained through producing 1.5M spacecraft modules for RD1 and 2M spacecraft modules for RD2. In general, RD2 is more expensive than RD1 because more mass is involved; five RD2 systems are needed to generate roughly the same amount of power as one RD1 system.

Figure 2 shows the comparison of the lifecycle baseline assessment to terrestrial renewable electricity production technologies, whose costs range from 0.02-0.05 $/kWh. The RD1 LCOE and RD2 LCOE are 12-31 and 32-80 times higher, respectively, than the 2050 projections for terrestrial alternatives. Therefore, our baseline analysis of SBSP designs does not return cost competitive results relative to terrestrial alternatives. For comparison, the average energy cost of a U.S. household in August 2022 was 0.167 $/kWh.[9]

---

[7] Due to the size and mass of the representative SBSP designs, for purposes of this study, we used available data from Space Exploration Technologies Corporation's (SpaceX's) Starship launch vehicle. because at this time it is anticipated to be the largest super heavy launch vehicle with data available. It is important to note that multiple specifications for this vehicle are planned, and cover a range of payload capacities, fuel capacities, and more. The study's use of data from Starship does not indicate any endorsement by NASA.

[8] Blue Origin Fed'n, LLC; Dynetics, Inc.-A Leidos Co., B-419783 et al., July 30, 2021, 2021 CPD ¶ 265 at 27 n.13

[9] U.S. Bureau of Labor Statistics (2023), Average energy prices for the United States, https://www.bls.gov/regions/midwest/data/averageenergyprices_selectedareas_table.htm.

The baseline lifecycle GHG emissions intensity for RD1 is 26 $gCO_2eq./kWh$ and for RD2 is 40 $gCO_2eq./kWh$. For comparison, the U.S. electric grid in 2021 produced an average of 385 gCO_2/kWh.[10] Launch is the largest driver and leads to 64% and 72% of the GHG emissions for RD1 and RD2, respectively. GHG emissions intensity for both RD1 and RD2 fall within the range of GHG emissions intensities (13-43 $gCO_2eq./kWh$) for terrestrial renewable electricity production technologies. For comparison, the GHG emissions intensities of coal and natural gas are 486 $gCO_2eq./kWh$ and 1001 $gCO_2eq./kWh$, respectively.[11] RD1 and RD2 emissions intensities do not include upper atmosphere effects of launch emissions, which are assumed to be worse than producing the same emissions on the surface of the Earth, and still under study by NASA and the academic community.[12] Our baseline analysis indicates our SBSP designs may have similar lifecycle GHG emissions intensities to those of terrestrial alternatives, pending further studies launch emission effects in the upper atmosphere.

Sensitivity Analyses

We conducted sensitivity analyses on the assumptions that drive the lifecycle cost and GHG emissions intensity to evaluate what conditions could allow RD1 and RD2 to be cost competitive (Figure 3). We varied the following input parameters one at a time to assess their individual impact on lifecycle cost and emissions: launch costs, first unit manufacturing costs, manufacturing learning curves, hardware lifetime, solar cell efficiency, and orbital transfer methods. Lower launch costs or use of electric propulsion to transfer mass from LEO to GEO each resulted in the most significant reduction of LCOE to about 0.20 $/kWh for RD1 and to about 0.50 $/kWh for RD2. This decrease is not enough to make the representative designs cost competitive with terrestrial alternatives.

Cost competitiveness can be achieved by varying multiple assumptions (Table 1) at the same time to provide a combination of cost and capability improvements beyond the advances already assumed in the baseline assessment. This favorable combination reduces the LCOE to 0.03 $/kWh for RD1 and 0.08 $/kWh for RD2, figures that are competitive with terrestrial alternatives. This combination also reduces the GHG emissions intensities (3.78 $gCO_2eq./kWh$ for RD1 and 4.33 $gCO_2eq./kWh$ for RD2) to values less than nuclear and wind-without-storage technologies.

---

[10] EIA, (2022, November 25), How much carbon dioxide is produced per kilowatthour of U.S. electricity generation? https://www.eia.gov/tools/faqs/faq.php?id=74&t=11.

[11] Ibid.

[12] National Oceanic and Atmospheric Administration (2022, June 22), Projected increase in space travel may damage ozone layer, https://research.noaa.gov/2022/06/21/projected-increase-in-space-travel-may-damage-ozone-layer/.

The following combination of revised assumptions yields SBSP solutions that are cost competitive with terrestrial alternatives, with lower GHG emissions:

- lower launch cost: $50M per launch, or $500/kg; $425/kg with 15% block discount
- electric propulsion orbital transfer from LEO to GEO
- extended hardware lifetimes: 15 years
- cheaper servicer and debris removal vehicles: $100M and $50M, respectively
- efficient manufacturing at scale: learning curves of 85% and below

Our sensitivity analyses highlight the need for advances across a wide range of SBSP enabling capabilities.

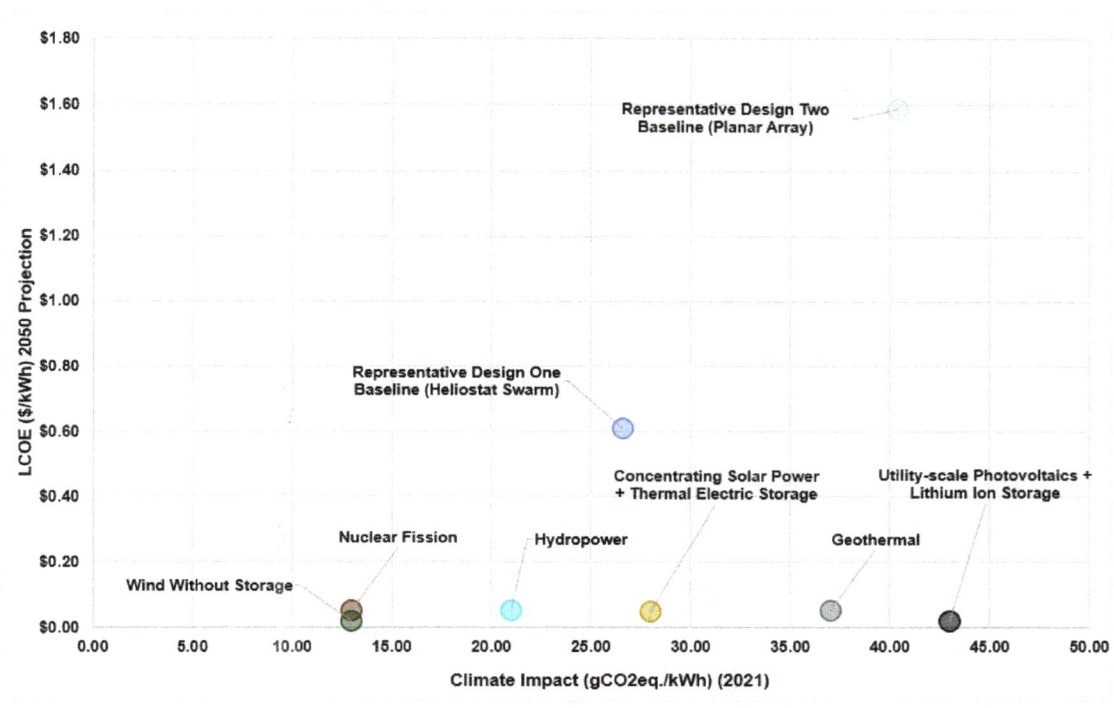

Figure 2. Comparison of SBSP Systems Cost ($FY22) and GHG Emissions Baseline Assessment with Terrestrial Alternatives

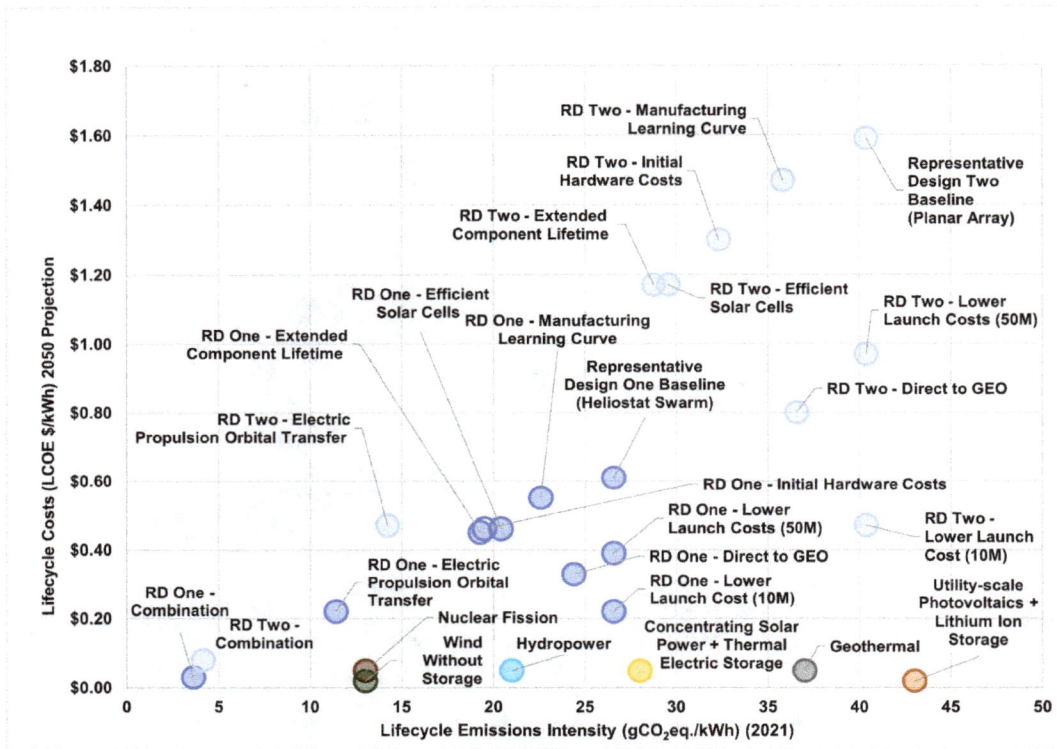

Figure 3. SBSP Systems Cost ($FY22) and GHG Emissions Reduction Sensitivities Results

Question 2: What role, if any, should NASA have?

Question 1 provided a model for understanding the biggest cost drivers for SBSP: launch and manufacturing. To understand NASA's potential role, the study qualitatively assessed challenges and opportunities for SBSP development. We reviewed technological, regulatory, and policy challenges, as well as technological and economic opportunities. The review found that SBSP enabling technologies have broad applicability to a wide suite of future NASA mission needs, from power beaming on the Moon, to autonomous operations for science and human exploration, to lightweight materials. NASA currently funds research and development activities in each of these areas, though some areas receive significantly more funding: In-space servicing, assembly and manufacturing received ~$280M in FY22, autonomy received ~$244M in FY22, while wireless power transmission investments are today limited to concept studies (<$1M).[13]

This study provides two main categories of options for NASA leadership to consider without making a specific recommendation:

1. **Undirected organic development**: NASA is working on almost all the enabling technologies for SBSP and may want to focus only on its own current and planned missions' needs, limiting further involvement, upon request, to supporting U.S. organizations pursuing SBSP and maintaining awareness of SBSP advances around world. NASA could fund these areas without adding SBSP as a separate line item in its budget. That said, further study of potential benefits of SBSP to planned missions is warranted.

2. **Pursue partnership opportunities to advance SBSP**: NASA may find mutually beneficial returns from supporting external SBSP development given the relevance of enabling technologies to other agency missions. Moreover, these technologies, from autonomous operations to wireless power transmission, have many use-cases beyond NASA missions, and are being pursued by a broad set of public and private actors for many non-SBSP applications.

This study also provides follow-on study recommendations regardless of option choice, including:

1. Building on the first order analysis, study cutting-edge SBSP systems using the most rigorous lifecycle emissions and cost assessments as performed by NREL on other electricity production technologies.

2. Perform a technical design trade evaluation of SBSP technologies for NASA mission applications, such as energy infrastructure on the Moon.

---

[13] Brandon, E. (2019, April 10). Power Beaming for Long Life Venus Surface Missions. Retrieved June 2023, from https://www.nasa.gov/directorates/spacetech/niac/2019_Phase_I_Phase_II/Power_Beaming/ and

Lubin, P. (2021, April 2). Moonbeam-Beamed Lunar Power. Retrieved from https://www.nasa.gov/directorates/spacetech/strg/lustr/2020/Moonbeam_Beamed_Lunar_Power/.

Conclusion

We performed a first-order lifecycle study of two representative SBSP designs for 2 GW utility-scale power generation that are presumed to begin operating in 2050 to determine 1) the conditions under which SBSP would be a competitive option to achieving net zero GHG emissions; and 2) assuming SBSP can be competitive, the role, if any, NASA could play in its development.

We assumed baseline capabilities to develop, assemble, operate, maintain, and dispose of the SBSP systems are a mix of capabilities that are above, below, or comparable to capabilities demonstrated to date. We then compared the LCOE and lifecycle GHG emission intensity of the SBSP designs to terrestrial renewable electricity production technologies. Our findings indicate the SBSP designs may produce lifecycle GHG emissions per unit of electricity that are comparable to terrestrial alternatives, pending further studies of upper atmosphere effect of launch emissions. We find the SBSP designs are more expensive than terrestrial alternatives and may have lifecycle costs per unit of electricity that are 12-80 times higher. However, cost competitiveness may be achieved through a favorable combination of cost and performance improvements related to launch and manufacturing beyond the advancements assumed in the baseline assessment.

NASA is developing technologies and capabilities to meet its future mission needs, such as in-space servicing, assembly, and manufacturing (ISAM) and autonomy, which are enablers for SBSP. NASA could maintain its focus on core Agency missions and technologies, while documenting their relevance to SBSP. NASA may also enhance coordination with U.S. and international partners on technology development with relevance to SBSP. We recommend regular reviews of global SBSP developments and focused analyses of SBSP designs that may enable NASA's core missions.

Table of Contents

1.0 Introduction

This report describes for NASA senior-level consideration the relative costs and greenhouse gas (GHG) emissions of space-based solar power (SBSP) systems to assess whether SBSP is a feasible option for achieving net-zero GHG emissions compared to alternative renewable sources of electricity production. Our assessment considered two reference SBSP system designs operating in geostationary orbit (GEO) – the lower technology readiness level (TRL) Innovative Heliostat Swarm, (hereafter referred to as Representative Design One, or RD1) and the higher TRL Mature Planar Array (Representative Design Two, or RD2) – and compared costs for their development, assembly, operation, maintenance, and disposal. We also compared the relative GHG emissions of each system by conducting material decompositions for an Economic Input Output – Life Cycle Assessment (EIO-LCA).

1.1 Background

In response to climate change, organizations around the world are pursuing a range of policies called net zero. According to the United Nations (UN), "net zero means cutting greenhouse gas emissions to as close to zero as possible, with any remaining emissions re-absorbed from the atmosphere, by oceans and forests for instance." There is growing U.S. and international policy and legislation on net zero. As of 2021 over 70 countries had set net-zero targets (United Nations, 2023). The U.S. submitted a long-term strategy to the UN in November 2021, officially committing to net zero emissions by 2050 at the latest (United States Department of State and the United States Executive Office of the President, 2021).

The electric power sector accounted for 25% of U.S. GHG emissions in 2020, according to the U.S. Environmental Protection Agency (EPA), as shown in Figure 2 (EPA, 2023). The sector encompasses the generation, transmission, and distribution of electricity. Carbon dioxide (CO_2) makes up 80% of GHG from the U.S. electricity sector. According to the U.S. Energy Information Administration (EIA), CO_2 emissions by the U.S. electric power sector in 2021 were about **1,545 million metric tons** (MMmt), or about 31% of the 4,970 MMmt of total U.S. energy-related CO_2- emissions (EIA, 2023). These emissions primarily result from electricity generation using coal and natural gas, which are non-renewable energy sources (see Figure 2 inset). In 2021, 40% of U.S. electricity production came from renewable and nuclear sources as shown in Figure 3. The International Energy Agency estimates that to reach net-zero, the world will need to reduce its use of fossil fuels from 80% of the total today to slightly over 20% by 2050 (Bouckaert, et al., 2021). However, the EIA projects that by 2050, 44% of U.S. electricity will still come from fossil fuels (EIA, 2022).

1

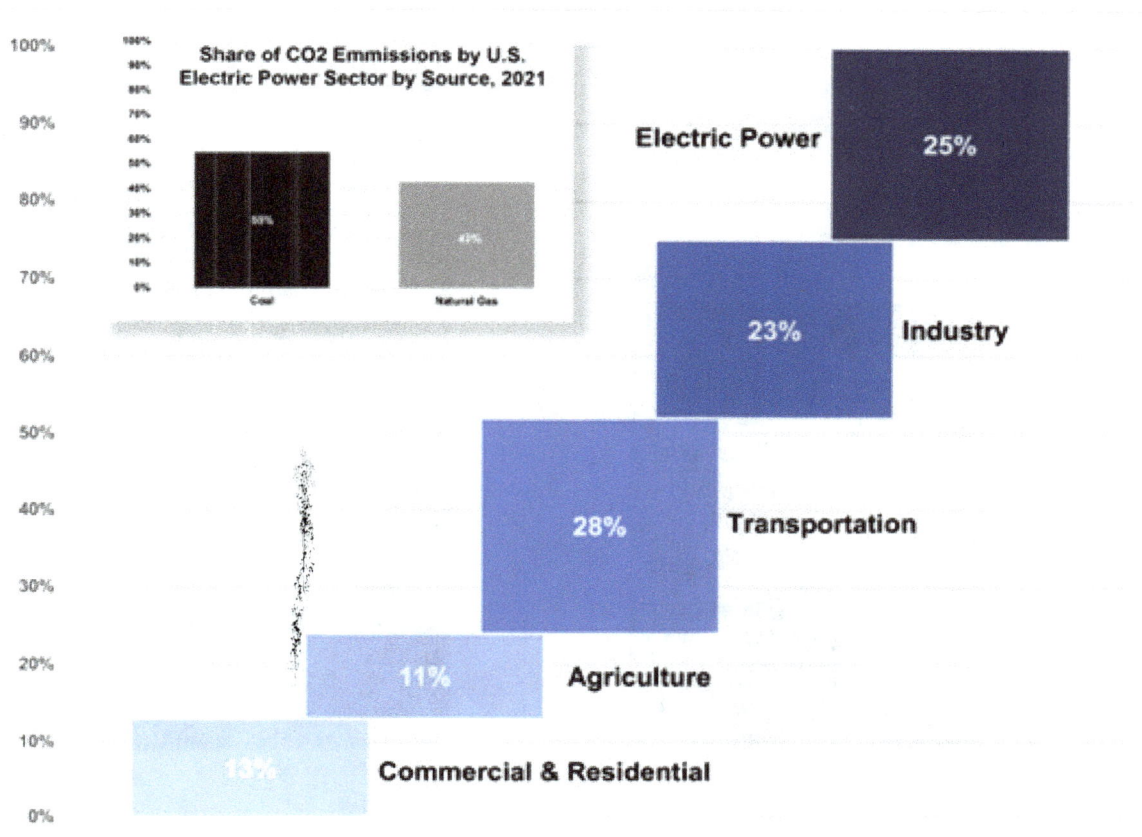

Figure 3. Share of U.S. Greenhouse Gas Emission by Economic Sector in **2020**, *(EPA, 2023). Inset. Share of U.S. CO_2 Emissions from Electric Power by Technology in* **2022** *(EIA, 2023).*

Some experts have noted that SBSP is a renewable energy alternative that could contribute to net-zero goals, though SBSP is not featured in any of the net zero pathways considered by the most recent International Panel on Climate Change (P.R. Shukla, 2022). An SBSP system collects solar energy in space, converts that to microwave or optical laser energy, and transmits that energy to the Earth. A ground station receives the energy, converts it to electricity, and delivers it to the power grid for use. The rate and intensity of worldwide research into SBSP has seen significant growth: The number of publications on the topic nearly doubled from 2018 to 2022, with most of the research concentrated in China, the U.S., the European Union (EU), Japan, and Russia (NASA Library, 2023).

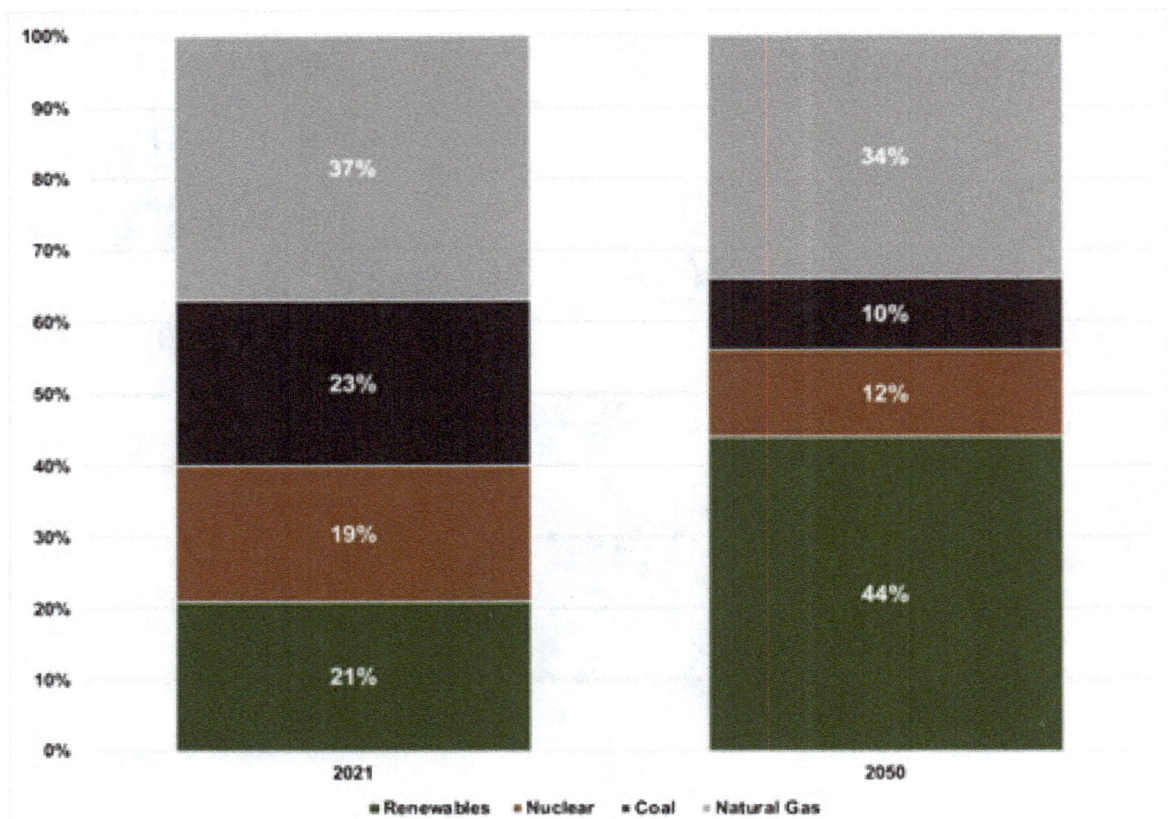

*Figure 4. U.S. Electricity Generation by Technology. [Left] Share of U.S. Electricity Generation by Technology in **2021**, (EIA, 2022). [Right] Projected share of U.S. Electricity Generation by Technology in **2050**, (EIA, 2022).*

Many countries have ongoing SBSP studies, design concepts, and technology development activities including the U.S., the United Kingdom, the EU, Japan, China, South Korea, and Australia. In the U.S., for example, the California Institute of Technology (Caltech) completed the first successful electricity beaming demonstration from space to ground in June 2023 (Caltech, 2023). In general, this work is funded and conducted by academic, commercial, and government communities motivated by economic development, net zero, and national policy goals. Figure 4 displays a map of current international SBSP activity. The SBSP concepts described in this report focus on civil applications of SBSP to deliver electricity to the power grid. SBSP is being pursued by different organizations for different use-cases. The resulting national benefits could extend beyond immediate fiscal returns or near-term GHG emissions reductions.

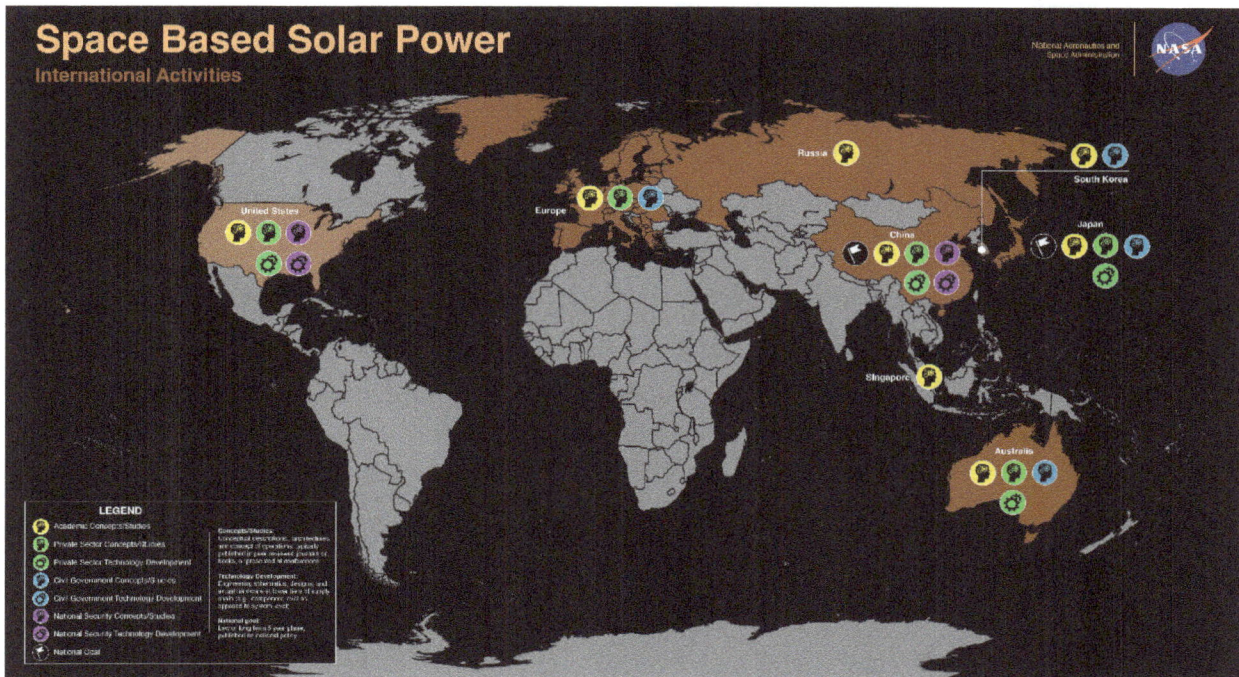

Figure 5. International SBSP Activities. SBSP studies, design concepts, and technology developments are funded around the world for economic development, net-zero goals, and national goals. Countries with non-space-based power beaming efforts are not included.

1.2 Study Questions

The idea of SBSP is not new to NASA, which conducted feasibility studies first in the 1970s (NASA & DoE, 1980) and again in the 1990s (Mankins, A fresh look at space solar power: New architectures, concepts and technologies, 1997). These studies found it prohibitively expensive to develop, launch and assemble, operate, maintain, and dispose of SBSP systems ($1T estimate in then-year dollars for an SBSP technology demonstration in the 1970s and $250B estimate in then-year dollars for the first commercial kilowatt (kW) of power in the 1990s). The context of SBSP development has changed significantly in the last three decades, however, prompting this study. Public and private actors across the international community are motivated to develop SBSP for economic development, net-zero goals, and global leadership.

The study seeks to answer two questions:

- Under what conditions would SBSP be a competitive option for achieving net zero GHG emissions compared to alternatives?

- If SBSP can be competitive, what role, if any, could NASA have in its development?

Alternative renewable electricity production technologies that we compared to SBSP include nuclear fission, geothermal, hydroelectric, utility-scale solar photovoltaics with storage, and land wind without storage. The study includes nuclear power even though it is not usually grouped with renewables because it is considered "non-emitting" by the EPA (EPA, 2023).

Given increasing investment and attention to SBSP worldwide, this study is intended to inform NASA decision-making regarding any potential Agency role in SBSP development. Therefore, we present options for consideration for senior leaders.

2.0 Methodology Overview

To determine the feasibility of SBSP we estimated the cradle-to-grave costs and GHG emissions of two system designs: RD1 (Innovative Heliostat Swarm) and RD2 (Mature Planar Array) based on existing concepts with updated technology assumptions on mass, efficiency, and launch capacity. The lifecycle cost estimates were used to calculate the levelized cost of electricity (LCOE) for each system for comparison to terrestrial renewable alternatives using data from the National Renewable Energy Laboratory (NREL). GHG emissions were estimated using a hybrid mass- and spend-based Economic Input Output-Life Cycle Analysis (EIO-LCA) and compared to terrestrial alternatives using NREL data.

Limitations of the study include:

- This is a first-order assessment of notional systems: Cost and GHG emissions estimates are not exhaustive, and outputs are heavily influenced by assumptions about a technology with no historical data points built and operated in an unknown future.
- The reference designs assessed are not representative of all proposed SBSP architectures: For example, systems in lower orbits have been proposed but are not assessed in this study.

For a detailed methodology, broken out by each step of the analysis and including results, please see Appendix B.

2.1 Cost Estimations

To estimate costs, we: 1) defined SBSP systems based on six key functions, 2) divided the SBSP system lifecycle into five concept of operations (ConOps) phases, generating cost estimates for each phase, and 3) used ConOps costs to determine the LCOE for comparison to other renewable energy technologies.

2.1.1 Functional Decomposition of SBSP Systems

Generating electricity using SBSP systems involves six functions: **collect** solar energy in space, **convert (in space)** energy to microwave or optical energy, **transmit** that energy to Earth, **receive** the transmitted energy at one or more ground stations, **convert (ground)** that energy to electricity, and **deliver** electricity to the grid for consumption or to batteries for storage.

This study assesses two representative SBSP designs: 1) the RD1 (Innovative Heliostat Swarm) concept, which uses a concentrator to improve its capacity factor, and 2) the RD2 (Mature Planar Array) concept, a less complex system that collects solar energy on one side and emits this energy as microwaves on the other. Figure 5 describes each reference design in terms of the six key functions; more detailed information on each concept is presented in Appendix A.

2.1.2 Concept of Operations

The ConOps for each reference design is broken into five lifecycle phases: develop, assemble, operate, maintain, and dispose. Including disposal or decommissioning of a system is a best practice when assessing its full lifecycle. Figures 6a and 6b provide a visual summary of each ConOps phase.

We estimate the cost of each SBSP reference designs by ConOps phase. Appendix A contains a detailed breakdown of each phase and all relevant parameters. Appendix B shows the mapping of ConOps phases to each functional step.

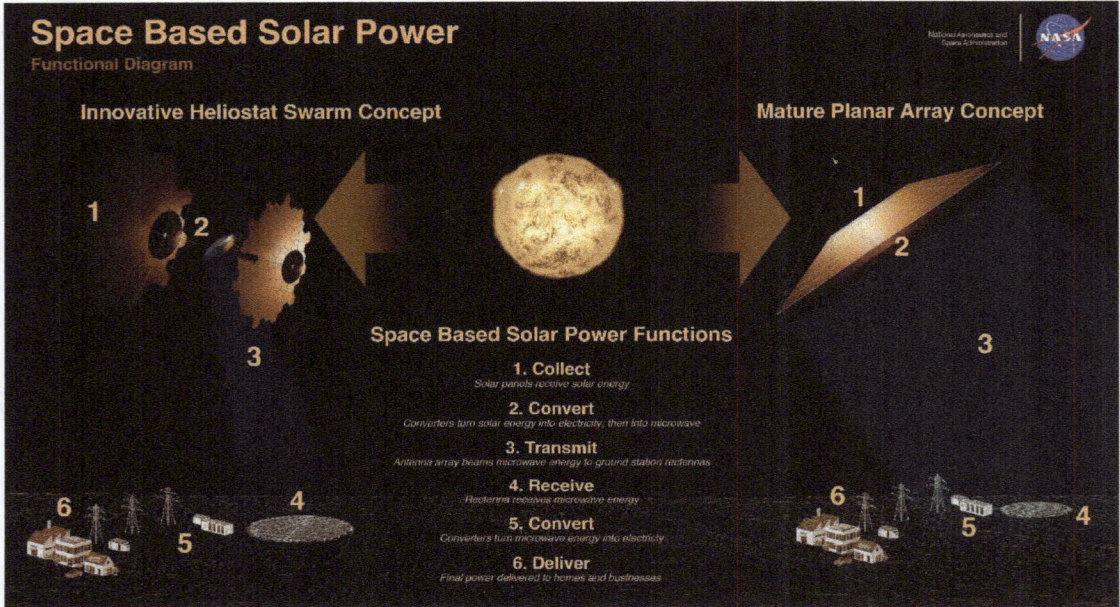

Figure 6a. Functional Decomposition of SBSP Design Reference Systems

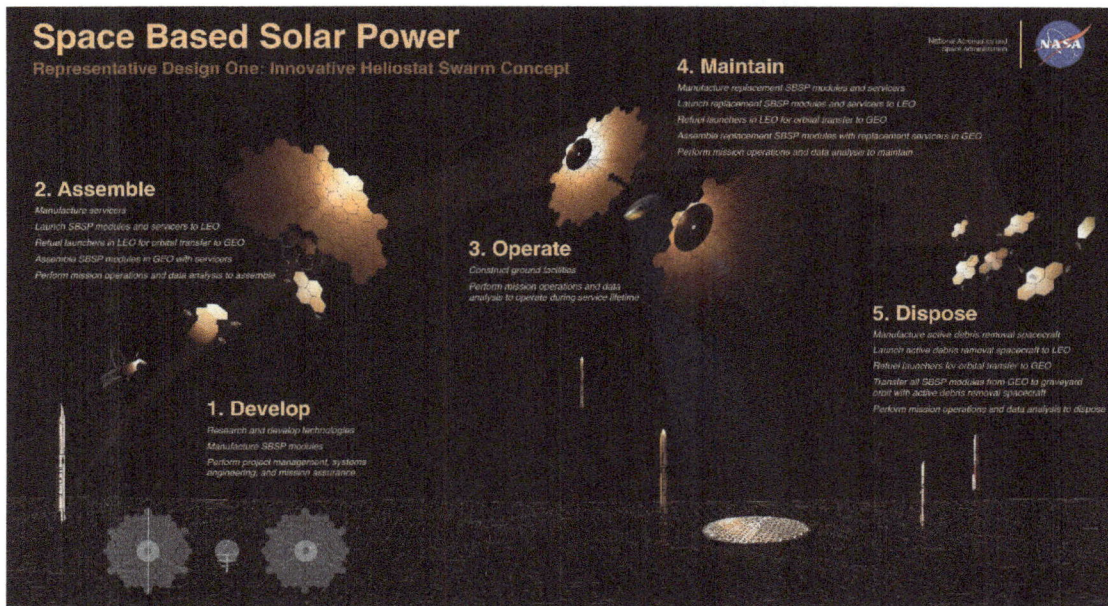

Figure 6b. ConOps for the Innovative Heliostat Swarm Design Reference One System

Figure 6c. ConOps for the Mature Planar Array Design Reference Two System

Baseline assumptions are derived from a mix of current and projected technologies and costs. Among the key assumptions in the baseline assessment is that Space Exploration Technologies Corporation's (SpaceX's) Starship launcher, which is currently in testing, will be commercially available at $100M per launch. We make no claim as to the reliability of this assumption given the early stages of Starship development, but rather make this assumption because on a per kilogram basis, this represents a similar decline in launch prices from today as has occurred in the past 10 years. We include a 15% block buy discount because of the very large number of launches required to deploy an SBSP system. We assume one payload-laden Starship in LEO requires refueling by 12 separate Starship propellant tankers to reach GEO (Blue Origin Fed'n, LLC; Dynetics, Inc.-A Leidos Co., 2021).[14] Based on subject matter experts (SME) input, we also assume 100 reuses of Starships used for refueling, and that payload-carrying Starships are single-use. SpaceX has been able to conduct two Falcon 9 launches in a week, so we assume this launch cadence for the Assemble ConOps phase. The Aerospace Corporation provided estimates of manufacturing learning curves, first-unit costs for each hardware element, component and system lifetimes, as well as module assembly time, all of which were reviewed, augmented, and incorporated by the study team. Finally, literature on SBSP concepts informed system specifications of the RD1 and RD2 designs. Estimated time to assemble a fully developed SBSP system in our baseline assessment is 7.4 years for RD1 and 12.6 years for RD2. Key input parameters are shown in Table 2. For a complete accounting of inputs, assumptions, and calculations, please see Appendix B.

2.1.3 Levelized Cost of Electricity

Using our cost estimates, we calculated the LCOE measured in $/kWh for each reference design to compare overall costs to other renewable electricity production technologies. LCOE is commonly used by the energy sector for comparative analyses. LCOE is calculated by estimating the lifecycle cost (using different units for different categories) of the SBSP system and dividing that by lifecycle kWh production by the system. Figure 7 provides an overview of the LCOE calculation.

---

[14] This estimate was derived using publicly available information about the initial human landing system for lunar exploration to be developed by SpaceX and modifying that based on the assumption that reaching GEO orbit would not require as many refueling launches as would reaching cislunar space. Blue Origins Fed'n, LLC; Dynetics, Inc.-A Leidos Co., B-419783 et al., July 30, 2021, 2021 CPD ¶ 265 at 27 n.13. This data was used because the Starship launch vehicle is anticipated to be the largest super heavy launch vehicle with available data and should not be construed as an endorsement by NASA. It is important to note that multiple specifications for this vehicle are planned, and cover a range of payload capacities, fuel capacities, and more.

Table 2. Key Input Parameters

Key Input Parameter	Value	Source and Rationale
Starship launch cost ($)	$100M, 15% block buy discount	2013-2022 36% launch cost decrease, Falcon Heavy cost of $1500/kg to 2030s. Scale of launches may offer discount.
Reuses of launch vehicle (does not include payload Starships traveling to GEO)	100	SpaceX states most components of Falcon 9 may be reused 100 times, but some elements must be replaced after 10 uses. New Glenn claims 25 reuses. 100 reuses = 4 times the state of practice.
Orbital transfer method	12 refuel launches in LEO 1 month refuel time 2 months travel to GEO	*Blue Origin Fed'n, LLC; Dynetics, Inc.-A Leidos Co., supra..* Does not consider cryogenic boiloff. First order delta-v.
Launches per year	104	Assuming two launches per week. Falcon 9 currently launching about 1.5x/week.
Manufacturing learning curves	85% for servicer 75% for modules 90% for debris vehicles	The Aerospace Corporation estimates based on aggregate manufacturing sector data.
First unit costs ($)	$1B for servicer $1M for SBSP modules $500M debris vehicles	The Aerospace Corporation estimates based on satellite and solar cell industry and OSAM-1 costs.
Solar cell efficiency	35%	NASA assessment of Smallsat technology.
Hardware lifetime (years)	10	The Aerospace Corporation estimates.
System lifetime	30	The Aerospace Corporation estimates.
Initial system upmass (kg), number of modules	5.9M (RD1), 1.46M 10M (RD2), 2M	Inferred from Mankins, Sasaki, Pellegrino.
Ground rectenna	6km diameter (RD1) 4km diameter, 5 sites (RD2)	The Aerospace Corporation assessed costs of analogs: solar power plants and large antenna arrays.
Operations cost	1.2M/month	Assumes autonomous operations capability. Significantly less than today ~500k/year per satellite.
Assembly time per module (minutes)	40 (RD1), 38 (RD2)	The Aerospace Corporation estimates based on Orbital Express and ISS analogs.

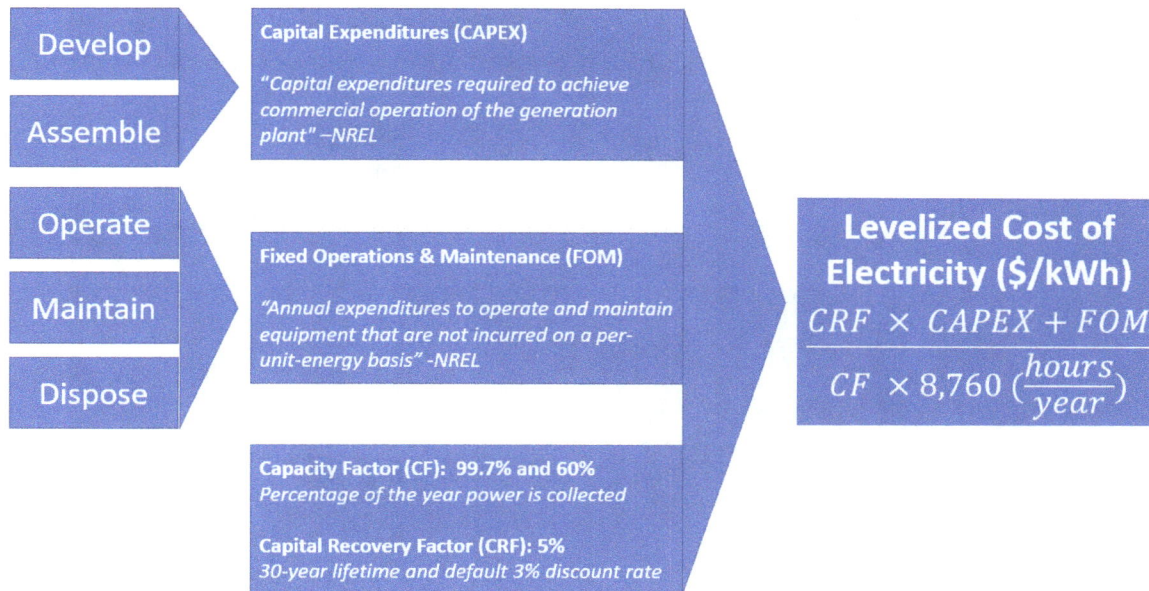

Figure 7. Cost Calculations for SBSP Systems. Refer to Figure 6 for ConOps Phase activities.

2.2 GHG Emissions Intensity

To estimate GHG emissions in carbon dioxide equivalents (CO_2eq), we 1) used material decompositions of each reference design, and 2) provided a comparison to other renewable energy technologies, drawing from NREL emission data (National Renewable Energy Laboratory, 2023).

We estimated the GHG emissions of each design in three steps:

1. Estimate the material composition of each design, in kilograms (kg) or square meters (m^2)

2. Cite authoritative sources on the emissions intensity of delivering components and materials, in $kgCO_2$eq per kg, m^2, or thousands of dollars (kUSD).

3. Estimate the lifecycle emissions intensity using a hybrid mass and spend-based Economic Input Output-Life Cycle Analysis (EIO-LCA).

EIO-LCA uses aggregate data on sectors of the U.S. economy to quantify the GHG emissions that can be attributed to specific sectors and activities. Our analysis uses the aggregated metrics provided by the International Aerospace Environmental Group (IAEG) (International Aerospace Environmental Group, 2023), including datasets from Carnegie Mellon University and the U.S. Department of Defense (DoD) on the GHG emissions intensity of activities measured in kilograms of carbon equivalents ($kgCO_2$eq.) per kUSD, kg, or m^2. This spend-based approach is used where material decomposition does not provide adequate coverage of post-processing and assembly work. It is important to remember that because the EIO-LCA model relies on aggregated economic transactions and their recorded GHG emissions, there is an assumed relationship between cost, efficiency, and reduced emissions, though it is possible to reduce costs without mitigating GHG emissions of manufacturing. The resulting GHG emissions estimates were then compared to other renewable energy technologies.

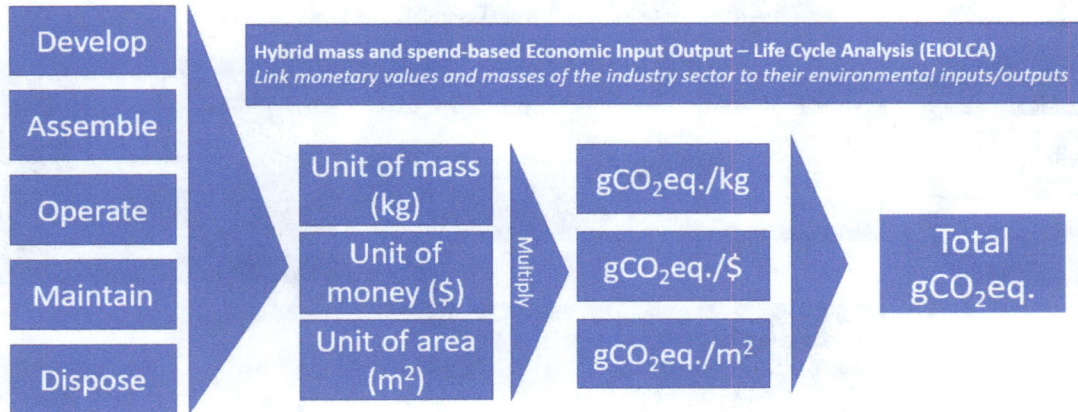

Figure 8. Calculations for SBSP GHG Emissions. Refer to Figure 6 for ConOps Phase activities.

Results of the initial baseline cost and emissions estimates were assessed to identify cost and climate drivers. We then conducted sensitivity analyses to determine the effect of incremental changes in these drivers.

3.0 Results

The study provides rough-order cost and GHG emissions estimates for the RD1 (Innovative Heliostat Swarm) and RD2 (Mature Planar Array) SBSP systems broken down by ConOps phase: **Develop, Assemble, Operate, Maintain, and Dispose**. Cost estimates for each ConOps phase by system are shown in Figure 9. For a detailed table of costs please see Appendix B. Both systems have a ~2 gigawatt (GW) capacity. The total estimated cost for each system is: RD1, $276B; and RD2, $434B.

For both systems **Maintain** comprises over 50% of the overall cost. **Assemble** costs comprise about 25% of total cost for both systems. The most impactful cost element is launch, representing 71% and 77% of total cost for RD1 and RD2, respectively.

Dispose, Develop and **Operate** are, in descending order, the next most expensive phases, but combined are less than **Assemble** for each reference design. The largest costs in **Develop** are for research and development (R&D), manufacturing and integration of all spacecraft hardware and systems, and program support services. Costs in **Operate** are primarily in the ground system; RD2 requires five ground rectennas where RD1 requires one. **Dispose** is unique in that it is the only ConOps phase where launch is included but is not the primary cost driver. For **Dispose**, the continuous operation of the Active Debris Removal (ADR) fleet for years is the largest cost.

The total GHG emissions for each system are: RD1, 14B $kgCO_2eq$.; and RD2, 21B $kgCO_2eq$. A breakdown of emissions by ConOps phase and cost elements is shown in Figure 10. As is the case with costs, **Maintain**, represents over half of each system's GHG emissions, with **Assemble** accounting for one quarter of the total. Similarly, the largest emissions across the SBSP lifecycle are attributable to the thousands of launches required, again highlighting launch's disproportionate impact on SBSP systems. Access to space comprises 64% and 72% of total emissions for RD1 and RD2, respectively.

In descending order, **Develop**, **Operate**, and **Dispose** produce the most emissions after **Maintain** and **Assemble**. The largest contributors in these segments is large-scale manufacturing of SBSP spacecraft, servicers, and launch vehicles. The activity with the next largest emissions is associated with the ground support infrastructure and staff, including R&D and operations.

Figure 11 depicts the LCOE and GHG emissions of the SBSP reference systems alongside other forms of renewable energy. The baseline costs, estimated using the assumptions described in the methodology overview, are significantly higher than those for current renewables, while GHG emissions are comparable. This remains true even when storage requirements to achieve a similar "power on demand" – also known as "baseload power" – for solar and wind are taken into account. Energy storage must be considered for solar and wind because they cannot deliver power consistently throughout the day or year. The LCOE for space-based systems is significantly higher as terrestrial systems do not face the high costs of launch and assembly in space, and this first-of-a-kind system's costs include R&D.

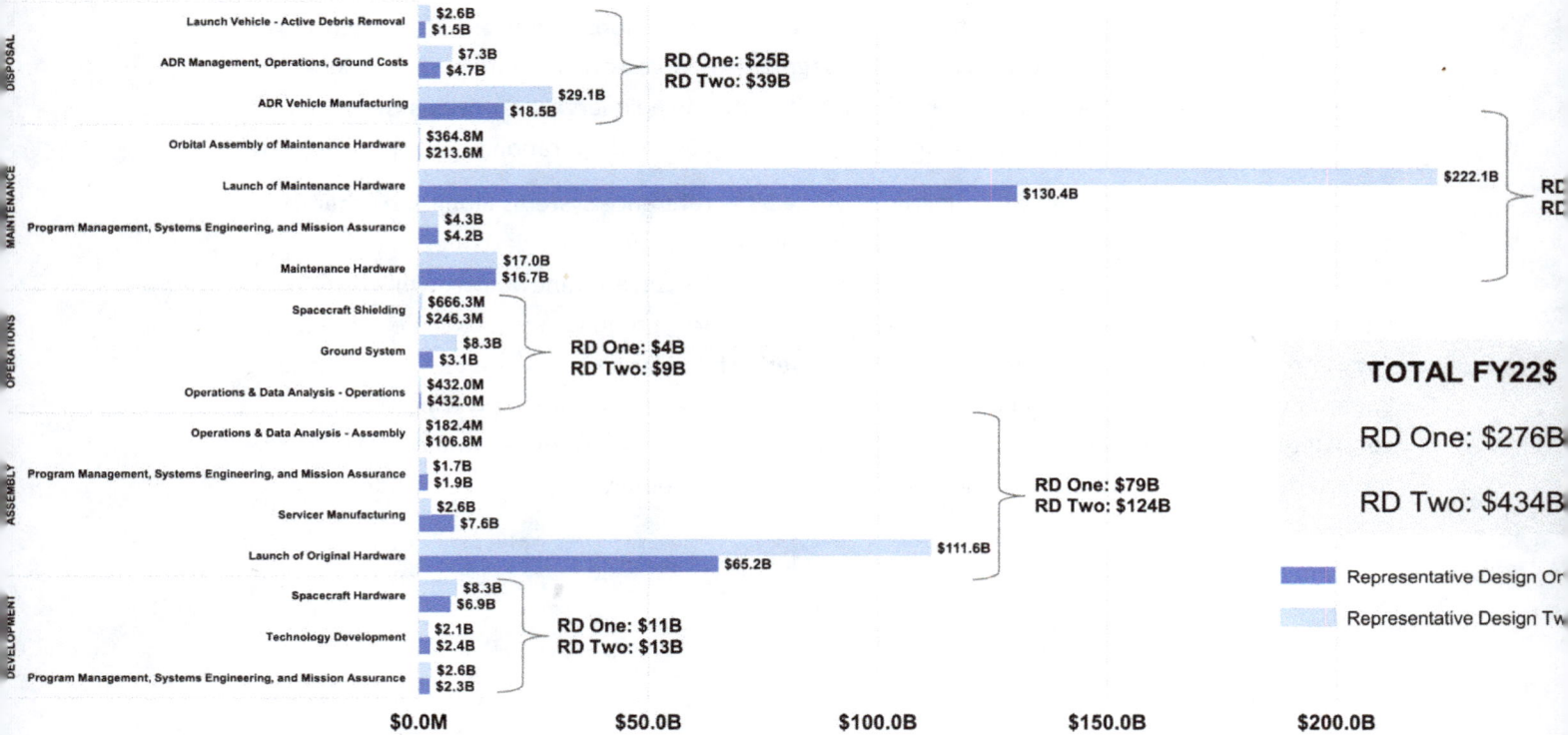

Figure 9. Cost Estimates Results of ConOps Phases

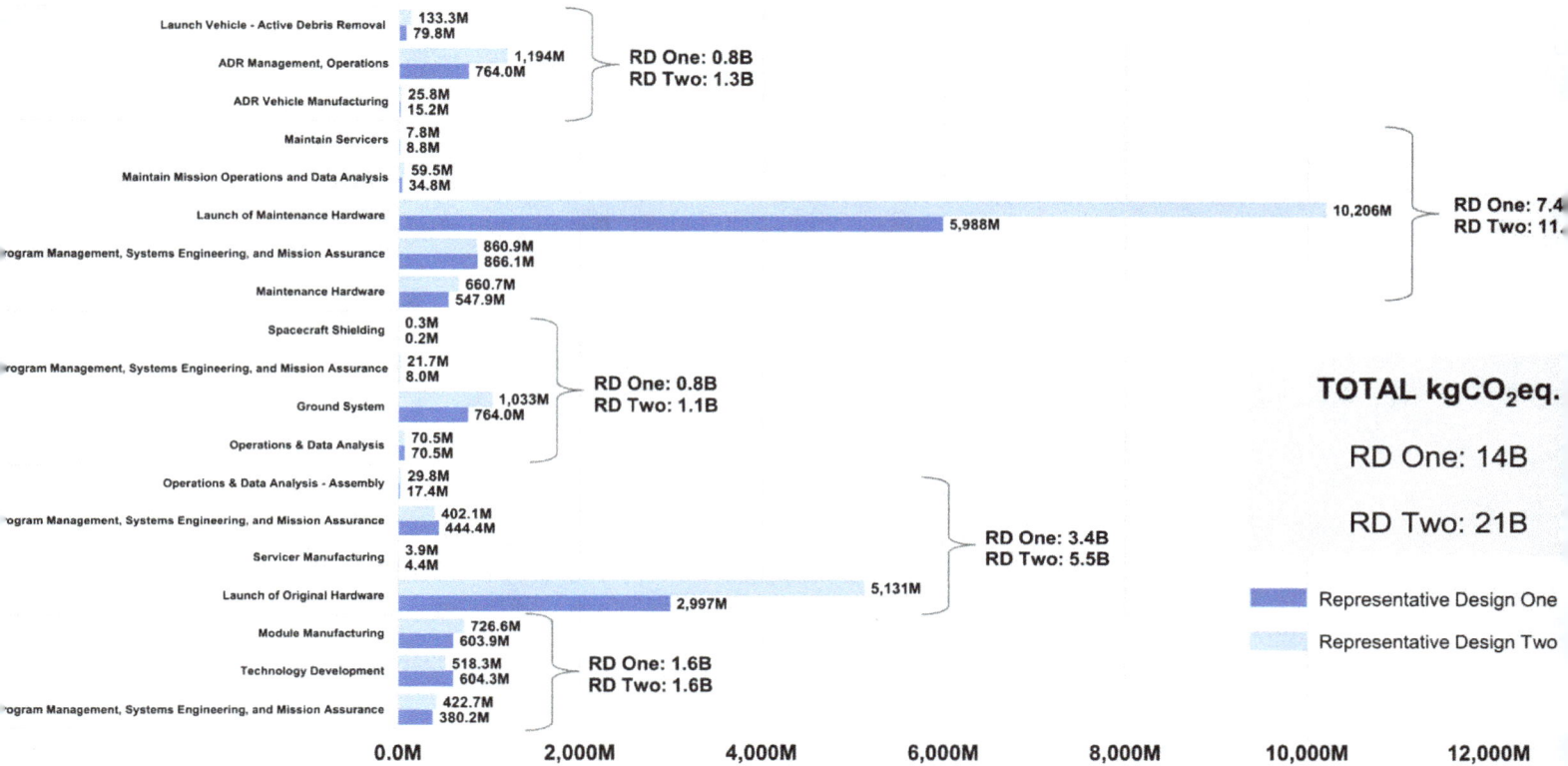

Figure 10. GHG Emissions Estimates Results of ConOps Phases

The chart shows GHG Emissions with the following values:

- Launch Vehicle - Active Debris Removal: 133.3M / 79.8M
- ADR Management, Operations: 1,194M / 764.0M
- ADR Vehicle Manufacturing: 25.8M / 15.2M
- Maintain Servicers: 7.8M / 8.8M
- Maintain Mission Operations and Data Analysis: 59.5M / 34.8M
- Launch of Maintenance Hardware: 10,206M / 5,988M
- Program Management, Systems Engineering, and Mission Assurance: 860.9M / 866.1M
- Maintenance Hardware: 660.7M / 547.9M
- Spacecraft Shielding: 0.3M / 0.2M
- Program Management, Systems Engineering, and Mission Assurance: 21.7M / 8.0M
- Ground System: 1,033M / 764.0M
- Operations & Data Analysis: 70.5M / 70.5M
- Operations & Data Analysis - Assembly: 29.8M / 17.4M
- Program Management, Systems Engineering, and Mission Assurance: 402.1M / 444.4M
- Servicer Manufacturing: 3.9M / 4.4M
- Launch of Original Hardware: 5,131M / 2,997M
- Module Manufacturing: 726.6M / 603.9M
- Technology Development: 518.3M / 604.3M
- Program Management, Systems Engineering, and Mission Assurance: 422.7M / 380.2M

Groupings:
- RD One: 0.8B / RD Two: 1.3B
- RD One: 7.4 / RD Two: 11.
- RD One: 0.8B / RD Two: 1.1B
- RD One: 3.4B / RD Two: 5.5B
- RD One: 1.6B / RD Two: 1.6B

TOTAL kgCO$_2$eq.

RD One: 14B

RD Two: 21B

Representative Design One
Representative Design Two

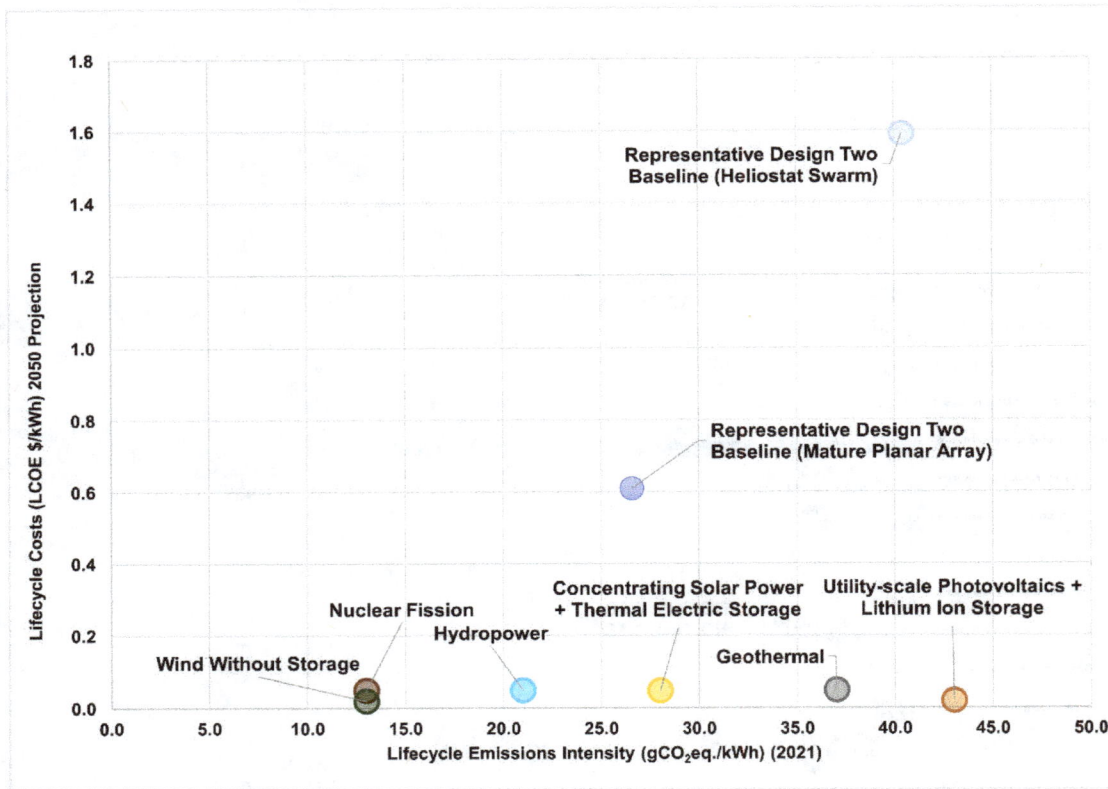

Figure 11. SBSP Systems (Baseline) and Other Renewables

3.1 Summary of SBSP System Costs and GHG Emissions

The total cost for each system is: RD1, $276B; and RD2, $434B. The total GHG emissions for each system are: RD1, 14B kgCO₂eq.; and RD2, 21B kgCO₂eq.

The combined on-orbit mass of the five RD2 satellites necessary to deliver 2 GW of power is nearly twice that of the 1-satellite RD1 system (1000 metric tons (MT) versus 585 MT), increasing its overall cost. The primary mass-based cost drivers are related to launch and disposal. The ground system costs for RD2 are also greater as it requires five ground rectennas to receive the same 2 GW of power. The same holds true for GHG emissions, as the increased number of launches leads to higher emissions relative to RD1.

The major cost and GHG emissions driver for both systems is launching and manufacturing millions of tons of hardware, including robotic servicers for on-orbit assembly.

Our baseline assessment assumes lower costs and better capability in some areas compared to today, and status quo in others, as discussed in Section 2, Methodology Overview. Recent analyses by Frazer Nash (Frazer Nash Consultancy, 2022) and Roland Berger (GmbH, Roland Berger, 2022) also make assumptions about the future state of technology and industry supply chains supporting notional utility-scale SBSP systems and have come out with much lower costs – LCOEs of $50-70/MWh – than our baseline numbers. Refer to the Methodology section or Appendix B for a full accounting of assumed capabilities and costs. Sensitivity analyses of the primary cost drivers show how SBSP costs could be reduced to levels more comparable with those found in other studies.

4.0 Sensitivity Analyses

Given that the largest cost drivers for both systems are launch and manufacturing of spacecraft and servicers, we conducted sensitivity analyses to assess the impact of reduced launch costs and improved manufacturing learning curves. First, we examined individual variables; then we conducted a multiple-variable sensitivity analysis. Details of the sensitivity analyses are provided below, and in greater detail in Appendix B. Figure 12 shows the difference in total cost from each sensitivity analysis to the baseline assessment.

4.1 Launch

4.1.1 Direct Launch to GEO

The baseline analysis assumes Starship launches 100 MT to LEO and is refueled on orbit to continue to GEO for system assembly. A Starship carrying 100 MT requires 12 refueling launches in LEO to continue to GEO, based on publicly available information about the initial human landing system to be developed by SpaceX and SME assessments. However, some Starship configurations are also capable of launching 21 MT direct to GEO. When launching directly to GEO, the number of launches drops from 2,321 to 863 for RD1 and from 3,960 to 1,470 for RD2. This decrease drops the LCOE of the RD1 and RD2 systems by 42% and 47%, respectively. This is assuming the cost per launch for a fully disposable or reusable Starship fleet is the same.

Direct launch to GEO yields only a marginal decrease in emissions, however, as the reduced number of launches is offset by the need to manufacture more vehicles because Starship vehicles launched to GEO cannot be reused. While all 863 direct-to-GEO missions for RD1 would be on single-use vehicles, for example, the 2,321 LEO launches required in the baseline assessment scenario could be carried out by just 199 individual vehicles. Direct launch to GEO saves about 2 and 4 gCO$_2$eq. per kWh – or 8% and 9% – for RD1 and RD2, respectively.

4.1.2 Reduced Launch Costs

Launches for assembly and maintenance are the biggest driver of SBSP system costs. Baseline launch costs for Starship are set at $100M/launch, the current price of Falcon Heavy. If that price is dropped to $50M/launch, the LCOE decreases by about 36% and 39% for RD1 and RD2, respectively. At $10M/launch the LCOE drops about 64% for the RD1 and 70% for the RD2. The decrease is higher for the planar array concept because requires more launches to deploy and assemble.

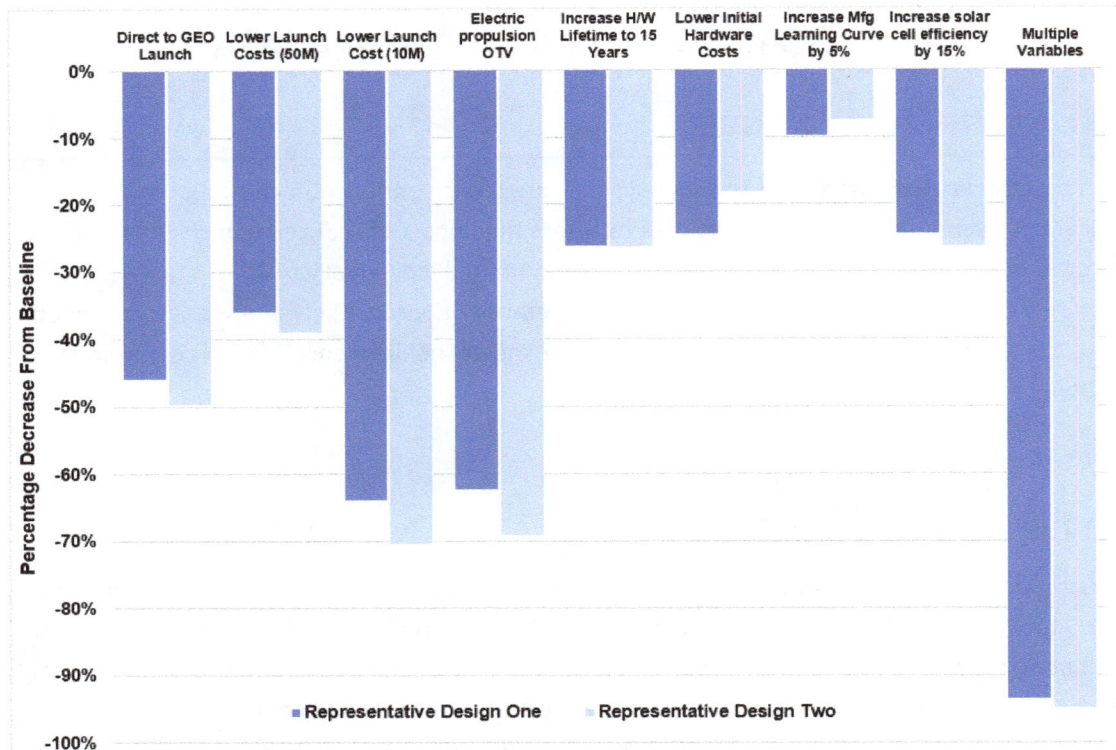

Figure 12. Percent Decrease in LCOE from Baseline

4.1.3 Electric Propulsion Orbital Transfer

Scholars have proposed using solar electric propulsion (EP) for orbital transfer from LEO to GEO as a cost saving approach. In this sensitivity analysis, 1720kg of propulsion system mass is allocated per 10,000kg of payload mass. We increased total hardware costs by 17.2% to account for the additional manufacturing cost of EP units. This approach takes advantage of launch vehicle reusability while eliminating refueling launches, lowering LCOE 63% for RD1 and 69% for RD2. Assuming EP specifications in line with NASA's NEXT-C electric ion thruster (NASA, 2023), travel time from LEO to GEO increases from one month to four.

Conducting orbital transfer with EP significantly reduces GHG emissions due to lower number of launches and fewer launch vehicles manufactured. The emissions decrease is about 54% for RD1 and 63% for RD2, potentially bringing those systems in line with terrestrial wind without storage.

19

4.1.4 Spacecraft Hardware Life

The baseline hardware lifetime is assumed to be 10 years, requiring two refurbishment cycles to maintain operations for a 30-year SBSP system lifetime. Extending hardware lifetime to 15 years halves the number of maintenance launches, decreasing the cost of both systems by 26%. GHG emissions are reduced by 27% and 29% from baseline for RD1 and RD2, respectively, due to the reduction in rocket manufacturing and launches.

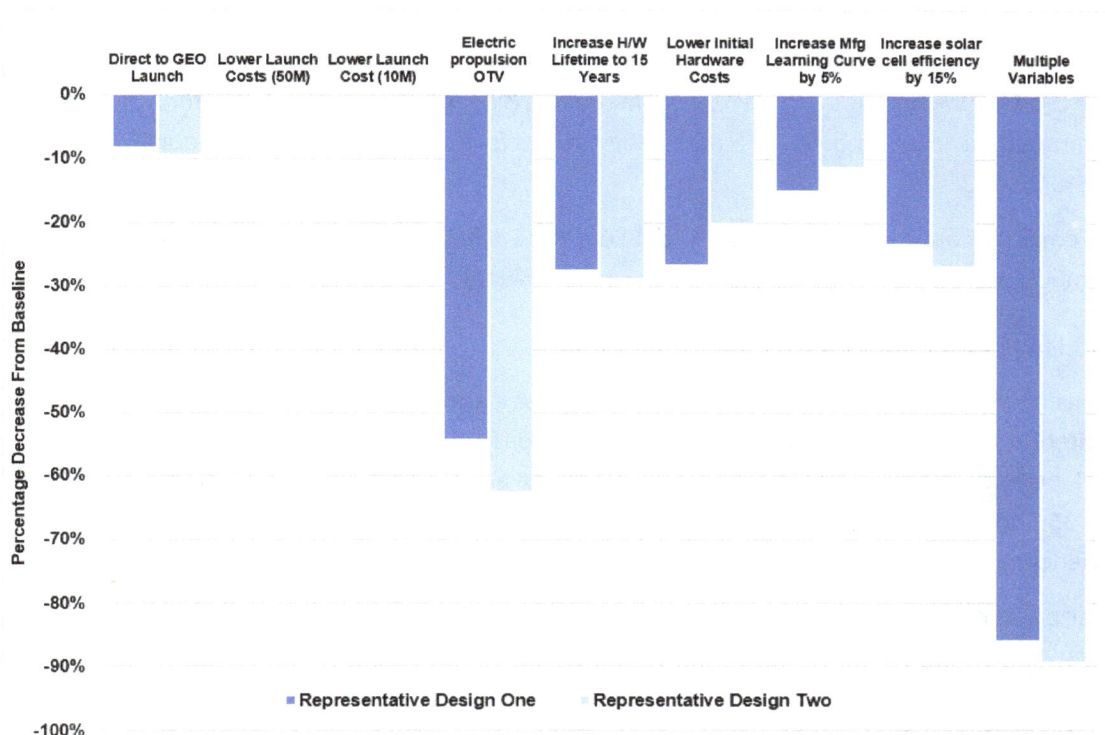

Figure 13. Percent Change in GHG Emissions Intensity from Baseline

4.2 Manufacturing

4.2.1 Initial Hardware Costs

The baseline cost estimates assume current hardware costs. Given the timeframe of SBSP development and deployment, hardware costs are likely to be lower, though we make no claim that the following changes represent any kind of prediction. For this sensitivity analysis, first-unit hardware costs were decreased 90%: The first SBSP system module's cost went from $1M to $100,000; the first servicer's cost from $1B to $100M; and the first ADR vehicle's cost from $500M to $50M. These values are in line with The Aerospace Corporation's survey of current commercial on-orbit servicers. These reductions in hardware costs reduced LCOE by about 25% for RD1 and 18% for RD2.

GHG emissions are decreased about 27% and 20% for RD1 and RD2, respectively, (Figure 13) because the EIO-LCA model assumes a relationship between cost, efficiency, and reduced emissions.

4.2.2 Learning Curve

Given the size of the systems and millions of modules required for each, manufacturing is one of the largest costs represented in the development segment of the ConOps. The baseline learning curve used is 75% for units of SBSP system hardware, 85% for servicers, and 90% for ADR vehicles. If we improve the learning curve by 5 percentage points for each, LCOE drops by about 10% for RD1 and 8% for RD2.

GHG emissions are reduced about 15% for RD1 and 11% for RD2 (Figure 13), again, because the EIO-LCA model assumes a relationship between cost, efficiency, and emissions.

4.2.3 Solar Cell Efficiency

Increasing the efficiency of solar cells decreases the size and mass of a space solar power system required to create the same output power. This decrease in size affects both hardware development and assembly costs. The LCOE reduction achieved by increasing solar cell efficiency from 35% to 50% is about a 25% for RD1 and 26% for RD2. The 50% figure represents the highest efficiency of terrestrial research cells tracked by NREL today (NREL, 2023).

The size and mass reduction leads to less manufacturing and fewer launches, cutting each system's assessed GHG emissions per kWh by 23% for RD1 and 27% for RD2. This sensitivity analysis does not consider the mass changes that would occur from use of different photovoltaic technologies. For example, silicon and perovskite cells are made of different materials, and manufactured differently.

4.3 Combining Sensitivities

Combining select conditions from the sensitivity analyses can lead to a cost-competitive SBSP system. The rationale for these choices is presented in the following section. In the "combined" sensitivity, we modified key input variables accordingly:

- Launch costs reduced from $100M or $1000/kg (with 15% block buy discount, $85M or 850/kg) to $50M or $500/kg (with 15% block buy discount, $42.5M or $425/kg)
- Solar cell efficiency increased from 35% to 50%
- Servicer and ADR first-unit cost reduced from $1B to $100M and $500M to $50M, respectively
- Learning curves improved by 5 percentage points across the board

- Hardware lifetime extended from 10 to 15 years
- Instead of the LEO-refueling deployment scenario, ConOps uses EP for orbital transfer, adding 17.2% more mass and cost to SBSP system hardware

Under these conditions, total system costs decrease by 95% and 93% for RD1 and RD2, respectively. GHG emissions decrease by nearly 86% and 89% for RD1 and RD2, respectively. These conditions would make SBSP systems highly competitive with any assessed terrestrial renewable electricity production technology's 2050 cost projections and 2021 emissions intensity. The effects of these changes to first-order cost and emissions estimates are shown in Table 3.

Table 3. Baseline versus Combined Sensitivities LCOE and GHG Emissions

	Sensitivity	LCOE ($/kWh)	GHG Emissions Intensity (gCO$_2$eq./kWh)
Baseline	RD1	0.61	26.58
	RD2	1.59	40.38
Combined Sensitivities	RD1	0.04	3.87
	RD2	0.08	4.33

4.4 Making SBSP Systems Competitive with Terrestrial Renewables

As seen in Figure 11, under the baseline assumptions neither SBSP system is cost competitive with other renewables. The costs of launch for in-space assembly significantly affect the LCOE of these systems, putting them far beyond the costs of terrestrial solutions. SBSB still would not be cost-competitive with alternative renewables even if access to space were free, assuming all else remains constant. A cost-competitive SBSP solution does emerge, however, if launch costs drop to $50M/launch, solar cells achieve 50% efficiency, costs for a commercial servicer decrease to $100M, learning curves improve by 5%, hardware lasts 15 years, and EP is used for the orbital transfer of payloads to GEO. Below, we briefly discuss the likelihood of these sensitivities:

- Launch cost: Launch costs per kilogram have decreased 36% over the last 10 years. If that continues, they could reach ~$60M per 100 MT (the assumed Starship payload capacity) in 2040. Note this cost is still 30 times greater than SpaceX's desired launch cost of $2M per Starship launch.

- Solar cell efficiency: According to NASA's assessment (NASA, 2022), the state of the practice of solar cell efficiency in space today is 33%, while the state of the art is 70% (based on theoretical limits of 6-junction solar cells in laboratories today). Similarly, while state of the practice solar cell efficiency on Earth today is about 20%, laboratory solar cells can reach efficiencies of 50% (Center for Sustainable Systems, 2022) (NREL, 2023). For comparison, the highest recorded solar cell efficiency in 2011 was 27.6% (Radiative efficiency of state-of-the-art photovoltaic cells, 2012). For every alternative solar cell technology however, a full design review would be needed to assess lifecycle effects on mass, cost, and emissions.

- Servicer cost: The Aerospace Corporation's survey of commercial offerings for in-space servicing, assembly, and manufacturing (ISAM) capabilities yielded a range of $75M-$750M, suggesting $100M may be possible in the next 15 years.

- Learning curve: Manufacturing learning curves in other industries range from 108% to 54%, though 80-82% is most common (Aerospace, 2018). For satellite manufacturing, The Aerospace Corporation estimated learning curves after reviewing manufacturing industries, including satellite manufacturing (Meisl & Morales, 1994).

- Hardware lifetime: The standard lifetime for GEO satellite hardware is currently 15 years.

- EP orbital transfers to GEO: These maneuvers have been conducted for over a decade (Boeing, 2012), and first order delta v calculations with state-of-the-art performance suggest feasibility at scales required, though the mass and configuration may require further assessment.

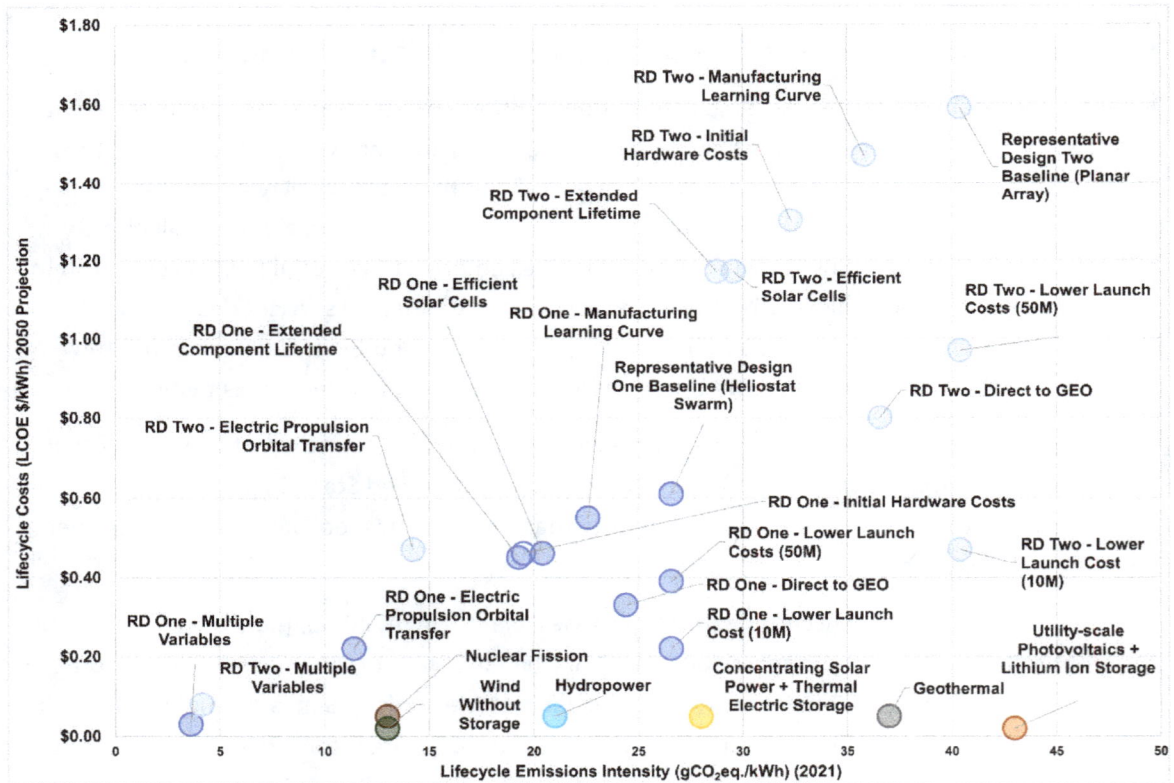

Figure 14. SBSP Systems Sensitivities Compared to Other Renewables

Using baseline assumptions, the SBSP systems assessed perform competitively on GHG emissions when compared to terrestrial forms of renewable electricity production. It is important to note, however, that the effects of burning rocket fuel in the upper atmosphere are not yet quantified by science (NOAA, 2022), but widely assumed to be much worse than burning these fuels on the ground. Given the large number of launches required, we would expect the measurable impacts of these systems to rise as we learn more. Manufacturing at scale for these massive systems and thousands of launches of rockets using methane-oxygen fuels produce the most GHG emissions.

Options to reduce GHG emissions:

1. Reducing the number of launches by using EP for orbital transfer, deploying the system at a lower altitude, or using non-combustive launch technologies (e.g. spin launch or magnetic launch) would reduce launch related emissions. Fewer launches also would reduce the environmental footprint associated with launch vehicle manufacturing, even taking reusability into account.
2. Although the baseline assessment assumes autonomy is fully developed, it is important to note the significant reduction in emissions this provides during assembly and operations.
3. On-orbit manufacturing of photovoltaic arrays and other components would remove manufacturing emissions that would otherwise occur on Earth.
4. Any technologies that reduce the mass of an SBSP system would reduce both the launches and manufacturing required to build, assemble, and maintain the system.

5.0 Challenges and Opportunities

Fielding either of the two SBSP reference designs analyzed in this report will require major capability advances in three key areas: 1) ISAM, 2) autonomous distributed systems, and 3) power beaming.

5.1 Challenges to Operational System Development

Across all technology areas, the single most distinguishing factor is the sheer scale of operations, mass, and coordination required to deploy systems larger and more massive than anything built in space before, except perhaps the combined mass and area of very large satellite constellations.

5.1.1 Large-scale ISAM Capability Challenges

ISAM capabilities required for developing and operating an SBSP system (ConOps Assemble and Maintain phases) are not currently available. There are few existing commercial capabilities for activities such as assembly, refueling, in-space manufacturing, and formation flying. The market for ISAM is unproven, with many proposed technologies as yet untested in space. The ISAM business case has development and operational risks that may limit private investment. Unless the Government steps in to support this nascent industry, services will be limited and expensive.

Without robust ISAM capabilities, the Assemble, Maintain, and Dispose ConOps phases will be significantly slower and more expensive, especially when assembling a solar panel collection area measuring 19km², like that of RD2.

5.1.2 Large-scale Autonomous Distributed Systems

Autonomy is making significant advances today, both terrestrially and in space. On-orbit capability, however, is closer to automation (executing preprogrammed actions) than autonomy (independent decision-making onboard a system), as seen in the collision avoidance capability of large-constellation satellites in orbit today. For SBSP, collaboration between multiple autonomous systems working across kilometers of space may be required to achieve cost targets. While this could theoretically be preprogrammed into automated commands instead of independent decision-making on satellites, either approach is still well beyond anything ever attempted. With today's technology, remote operators would be required for at least some operations, making assembly labor intensive and more time consuming than is assumed in this study's baseline assessment.

5.1.3 Power Beaming

Power beaming is a technical challenge that was not discussed as part of the sensitivity analyses. Power beaming from space was first demonstrated in June 2023 (Caltech, 2023), though at a scale that is orders of magnitude below what is baselined for the systems studied in this report or proposed elsewhere. Significant advances are needed before commercial-scale SBSP power-beaming systems would be technically feasible.

5.2 Challenges to Reducing System Costs

As discussed in Section 3, Results, the reference design systems are far more expensive than and may have comparable GHG emissions to terrestrial renewable alternatives. The sensitivity analysis section of this report showed that both costs and emissions may be reduced through advances in technology and access to space. However, many of these technologies are still in development as described in this section.

Every element of the SBSP ConOps has significant cost challenges. These challenges are associated with the sheer scale of the manufacturing effort as well as the number of launches required. They could be mitigated through manufacturing and launch efficiencies, or both.

5.2.1 Launch costs

High space transportation costs are the single most impactful cost barrier. Launch costs affect every in-space element, from SBSP system construction and maintenance to ISAM capabilities like refueling and assembly.

Analysis shows (BryceTech, 2022) that the average launch price per kilogram of payload dropped by 36% from 2013 to 2022. Our baseline assessment assumes launches begin around 2040 to assemble the SBSP system in orbit. If the current rate of decline continues – today's lowest price is $1,500/kg – launches would cost $615/kg in 2040, or $61.5M for the 100 MT each Starship delivers to LEO. This is still higher than the $50M figure we used for our multiple-variable sensitivity analysis.

As previously noted, the exact climate impact of burning rocket fuel in the upper atmosphere is unknown, but widely assumed to be worse than burning an equivalent amount of fuel on the ground (NOAA, 2022).

5.2.3 Manufacturing at scale

To lower the manufacturing cost of SBSP infrastructure, it is important to achieve economies of scale. Mass production has driven down the cost of mega-constellations, making the services offerings of Starlink (SpaceX, 2023) and OneWeb (OneWeb, 2023) price competitive for consumers. SpaceX reportedly produced 120 satellites and thousands of terminals per month during 2020, an unprecedented rate in the industry (Sheetz, SpaceX is manufacturing 120 Starlink internet satellites per month, 2020). In the SBSP Develop phase, to lower module manufacturing costs, scaled manufacturing will be required with significant upfront capital expenditure.

Despite significant advances in mass manufacturing of satellites, learning curves and costs have not come down to the level needed for the scale required to support SBSP system development. Findings from this report and supporting research suggest the industry needs to reach annual production of at least thousands of satellites to reach the learning curves described in this assessment. High hardware manufacturing costs therefore remain a barrier in the Develop phase.

5.2.4 Launch cadence

Launch cadence affects the cost and schedule of SBSP systems. The baseline assessment assumes two dedicated SBSP launches per week, or 104 per year. For comparison, *global* launches in 2022 set a new record with 186 (BryceTech, 2023). Currently there are no plans to expand launch capacity to the degree required to support our assumed SBSP launch cadence.

It is also important to note that while Starship provides far greater payload capacity than existing operational vehicles and is assumed in the baseline assessment to be somewhat cheaper per kilogram, it has unique ground system requirements that limit its launch capacity. To meet our baseline assumption of 104 launches per year, Starship would require five to 10 times its current ground support infrastructure, all dedicated to SBSP. This may not be possible given competing needs for Starship from other customers.

5.3 Regulatory and Other Challenges

There are non-technical issues with SBSP that may prove challenging, such as debris remediation, spectrum allocation, orbital slots, and security.

5.3.1 Active Debris Removal

SBSP is both vulnerable to space debris and a contributor to space debris. While some debris shielding is assumed in this study's assessment, it is unlikely to provide sufficient protection, and an SBSP system is unlikely to have sufficient warning and maneuverability to avoid collisions with space objects. This would lead to a risk of cascading debris creation and must be mitigated. Moreover, there likely are risks associated with moving a retired SBSP system into a super-GEO graveyard orbit. Moving the debris closer to Earth for disposal would also present risks and costs. The baseline assessment assumes disposal in a graveyard orbit, but this is not necessarily the best course of action. On-orbit industrial capabilities that could recycle debris are not available today, nor is their future availability assured.

5.3.2 Spectrum Allocation

Radio spectrum is a finite resource that requires coordination with the International Telecommunication Union (ITU) and is subject to both international and domestic regulation. SBSP transmissions from GEO may cause interference with other systems, especially those operating LEO. Uncertainties remain about what frequencies a future commercial-scale SBSP system will operate in and what technologies will be available to allocate popular frequency ranges to SBSP operations. There are additional concerns about the warehousing of, and competing interests in, radio frequencies. Research is required to better understand how SBSP transmissions might affect other satellites, particularly those operating in lower orbits.

5.3.3 Orbital Slot Allocation

The ITU also allocates orbital positions for GEO satellites. Just like spectrum, orbital slots are increasingly contested (Gangestad, 2017). Coordination is required, necessitating prior planning to secure orbital slots for SBSP missions. The larger the SBSP system, the greater the challenge.

5.3.4 Security

While not unique to SBSP systems, security concerns will have to be addressed. Space systems and critical infrastructure like power plants have security requirements, and an SBSP system would need to consider all of these risks. Cybersecurity may be of particular concern for highly autonomous and distributed systems.

5.4 Ongoing Improvements to SBSP Technology Needs

Most SBSP challenges and opportunities are shared by the global space sector. There are ongoing efforts to leverage emerging opportunities and industry trends to address many of these challenges. Below we discuss some representative examples.

5.4.1 ISAM

Large-scale ISAM capabilities are necessary to achieve the baseline scenario assumed in this study. ISAM capabilities not only enable the reference designs reviewed; they also reduce the cost of assembly, operations, and maintenance, while also supporting disposal and debris remediation.

In the U.S., the Government and private sector are working to advance ISAM technology. The Office of Science and Technology Policy's (OSTP) National ISAM Strategy and Implementation Plan guides interagency objectives (United States National Science and Technology Council, 2022). The baseline assessment servicer price and mass are derived from NASA's On-orbit Servicing Assembly and Manufacturing (OSAM)-1 mission and Northrop Grumman's Mission Extension Vehicle (MEV) because of the detail of publicly available information. Northrop Grumman (Northrop Grumman, 2023) provides in-orbit GEO satellite servicing using the MEV. Other commercial actors are also offering ISAM services. Servicer price in select sensitivity analyses is derived from alternative offerings. OrbitFab (OrbitFab, 2023) is offering hydrazine refueling services in GEO for $20M and KMI (Kall Morris Inc., 2023) has offered debris removal services at costs ranging from $4M to $62M. These prices are much lower than the baseline assumptions used in this study. Several startups are working on SBSP (Kirschner, 2023) concepts that would leverage ISAM technology. For example, in 2023, Orbital Composites and Virtus Solis announced a memorandum of understanding (MOU) for development of an SBSP system. Many more startups are working on ISAM specifically, as seen in the case of the U.S. Air Force Research Laboratory's SpaceWERX Orbital Prime program, which awarded ISAM-related contracts to 124 companies in 2022 (Holt, 2022).

5.4.2 Autonomous Distributed Systems

ISAM capabilities are, in most cases, leveraging autonomy at the system level. SBSP also requires a large number of satellites working together. Autonomous distributed systems and formation flying technologies that currently exist for military and aviation applications can be applied to support SBSP systems. NASA efforts including the Distributed Spacecraft Autonomy project (NASA, 2020), along with the deployment of very large satellite constellations, are advancing this critical SBSP technology.

5.4.3 Power Beaming

Power beaming technology is being pursued by many organizations globally. Like other activities benefiting SBSP, power beaming is not only being pursued for SBSP; power beaming on Earth has

28

been demonstrated many times and in-space power beaming was demonstrated by the Naval Research Lab (NRL) in March 2023 (Hamisevicz, 2023). Beaming from space to Earth was first demonstrated in June 2023 (Caltech 2023), at a very small experimental scale.

There is additional active U.S. federal research aimed at developing advanced power beaming capabilities. The Defense Advanced Research Projects Agency (DARPA, 2022) Persistent Optical Wireless Energy Relay (POWER) program is focused on power beaming for terrestrial applications, while the Air Force Research Laboratory (AFRL) is developing an in-space power beaming experiment. NASA's efforts are currently focused on research rather than technology development, examples being the NASA Innovative Advanced Concepts (NIAC) Low TRL concept for power beaming on Venus (Brandon, et al., 2020) and Lunar Surface Technology Research (LusTR) Beamed Lunar Power on the Moon concept (Lubin, 2021).

Other countries, such as Japan and South Korea, are working on power beaming. Japan (Kawahara, 2023) plans to test beaming energy from space to Earth in 2025, for example. International collaboration could speed up development of key SBSP technologies.

5.5 Ongoing Improvements to SBSP Economic Needs

It is important to note the significant overlap between technical and economic opportunities as they relate to SBSP. These technologies, while not directly required for SBSP operation, would make SBSP more cost effective.

5.5.1 Electric Propulsion Orbital Transfer

Using EP to transport SBSP payloads from LEO to GEO will significantly decrease the number of launches, in turn decreasing overall system cost as well as GHG emissions. EP for orbital transfers is a proven technology in use by GEO satellite operators seeking to reduce launch costs in exchange for a longer travel time (de Selding, 2013) (Werner, 2018). However, there is no history of moving so much mass from LEO to GEO and this may require EP capabilities that do not exist today.

Currently, several companies have small space tugs with EP, such as Spaceflight's Sherpa-LTE Orbital Transfer Vehicle (OTV), Starfish Space's Otter, Momentus' Vigoride, and Atomos Space's Proton and Quark OTVs. Exolaunch's Reliant eco-friendly space-tug vehicle has a "Pro" version that combines electric and green propulsion. Sherpa-LTE by 2022 had achieved its 300th on-orbit burn. These tugs are sized for satellites that have two orders of magnitude less mass than could be launched by a single Starship rocket. Hybrid electric and chemical OTVs were not explored in this study but could offer a faster solution that is not significantly worse for the environment than conventional chemical systems.

5.5.2 Alternative Launch

Some alternative launch technologies in development may lead to significant cost reductions. Multiple companies are researching kinetic launch, for example. SpinLaunch (SpinLaunch, n.d.) has a Space Act Agreement with NASA and Longshot Space (Longshot Space, n.d.) is performing work for the Air Force under its Small Business Innovation Research program. Given existing mass limitations, these approaches would only apply to modules and not servicers, and it remains to be seen whether kinetic launch will be successful and achieve promised cost reductions.

5.5.3 Mass Manufacturing

Advances in mass manufacturing would benefit SBSP, certainly the multi-million-module designs examined in this study. Mass manufacturing, as seen with Starlink and OneWeb satellites, may reduce space hardware production costs in general, potentially benefitting unrelated space manufacturing activities, including SBSP. While manufacturing does tend to improve with experience, and the scale of SBSP constitutes significant experience, we cannot assume that this will be the case until it happens.

5.5.4 Advanced Materials

Advances in materials could lead to significant mass reductions, whether from the material itself or new design optimizations. NASA has invested in several relevant projects, such as Lightweight Materials and Structures (NASA, 2020) and Superlightweight Aerospace Composites (NASA, 2022). Novel materials may be more expensive until production scales, which may limit their applicability.

5.6 Architecture Opportunities

This study did not assess the potential of novel architectures. However, improvements and new designs currently being studied may lead to lower costs or higher efficiency. These opportunities may change SBSP ConOps and introduce significant cost reduction. In the baseline estimate, launch costs represented one third and half of total costs for RD1 and RD2, respectively. Caltech has explored deployment of a medium Earth orbit (MEO) constellation of SBSP satellites (Marshall, Madonna, & Pellegrino, 2023). Incorporating Caltech's 2022 analysis of a MEO constellation's efficiency into this study's cost and climate model would provide a more complete picture of SBSP system efficacy.

6.0 Options for NASA to Consider

In this section, we present options for NASA senior leaders regarding the Agency's role in the worldwide acceleration of SBSP development activity. This data may provide decision-makers with a basis to reassess existing policy. The two options discussed below are not mutually exclusive; NASA may elect to pursue both paths.

Our research indicates NASA is developing technologies with broad applicability to a wide suite of future mission needs and enable SBSP as well. However, we view SBSP as a use case for these technologies, not a driver for NASA's development programs. We recommend that NASA stay abreast of outside SBSP developments and requirements as it matures the technologies needed for its missions. NASA could maintain its awareness in part by repeating this study at different scales of effort every three to five years.

6.1 Option 1: Undirected and Organic Development

NASA is currently developing technologies and capabilities that have applicability to SBSP, such as ISAM, autonomy for distributed systems, and power beaming, though these investments vary greatly. ISAM and autonomy have dedicated technology development programs while NASA-funded power beaming work today is limited to paper research. Continuing to invest in these capabilities, even while taking no new action, will make SBSP systems more technically feasible in the future. NASA should continue to monitor and maintain awareness of ongoing developments in SBSP. This option does not require any changes to NASA's budget allocations.

6.2 Option 2: Pursue Partnership Options to Advance SBSP

Our work has shown there are multiple entities in the U.S. and abroad pursuing capabilities for which SBSP is a use case. This presents partnership opportunities. Specifically, NASA could become an SBSP technology development partner with other U.S. Government agencies, industry, academia, or international organizations. NASA could focus such partnerships on technologies with high-impact potential for the Agency's missions.

We identified several scalable opportunities for NASA to enter into strategic partnerships based on the level of technology maturity and funding availability. The list of examples below is not intended to be comprehensive, nor are the identified opportunities mutually exclusive.

1. Support other U.S. Government efforts.
   a. AFRL and DARPA are advancing power beaming technologies.
   b. DoE has several relevant power and infrastructure programs.
2. Support international efforts.
   a. ESA, Japan, Australia, and South Korea are U.S. allies and have ongoing government investments to develop SBSP technology.
   b. NASA could support ESA's interest in developing SBSP for net zero.
3. Explore novel public-private partnership opportunities in SBSP.
   a. A joint NASA-DoE study is recommended as a first step to inform such public-private partnership opportunities.

b. Given the existing commercial interest and activity in developing the necessary technologies, NASA and DoE could jointly assemble a consortium to support and guide maturation of these technologies.

c. Given that many more commercial companies are developing SBSP-relevant technologies for non-SBSP applications, such as power beaming and solar cell advances, we recommend collaboration with DoE and industry.

d. NASA could hold challenges and workshops to improve designs and develop better understanding of the required development paths.

e. NASA could advocate for the U.S. Government to create public-private entity with NASA as one of several Government shareholders to guide the development and implementation of SBSP.

7.0 Conclusion and Recommended Further Study

Our first-order assessment has shown that two notional SBSP systems, using existing or near-term technology, are very expensive but may produce GHG emissions comparable to existing renewable electricity production technologies. Some major drivers of cost and GHG emissions for SBSP include launch, space hardware manufacturing, disposal of massive satellites, and in-space assembly of large systems. However, our sensitivity analyses demonstrated that there are ways to significantly drive down the cost and emissions of SBSP systems. Specific opportunities that could also benefit a wide range of future NASA missions include using EP for transfer to the desired orbit, significantly decreasing the cost of access to space, improving solar cell efficiency, and improving manufacturing learning curves. A combination of such improvements would make SBSP systems competitive with other renewable energy systems on both cost and GHG emissions metrics.

Our work identified several policy and technology challenges that would need to be addressed to advance SBSP. While these challenges are not unique to SBSP systems, the scale required to fully implement an SBSP system requires forethought. We therefore recommend that NASA conduct follow-on assessments on the following items:

1. More detailed technical evaluations of SBSP – SBSP would benefit from a more detailed analysis of: 1) lifecycle cost and GHG emissions, as performed by NREL on other electricity production technologies; and 2) exploration mission applications of SBSP in the form of a cost benefit analysis and a detailed NASA technical design trade evaluation. The latter may also consider emerging partnership efforts for in-situ resource utilization, such as NASA-funded work to produce solar cells on the Moon (NASA, 2023).

2. Regulatory Challenges – SBSP faces several regulatory hurdles, including spectrum allocation, orbital slots, and launch approvals, that will need to be addressed given the competition for these limited resources and the complexity of proposed SBSP systems.

32

Further analysis is required to understand how regulations could impact the feasibility of SBSP systems and how NASA should work with the relevant regulatory bodies.

3. Policy Challenges – Several policy implications could impact SBSP deployment. The most notable concern is the safety of power beaming. In addition, the anticipated number of launches for assembly and refueling introduce orbital debris concerns at a time when many nations are looking to reduce their orbital footprint. Cybersecurity is also a significant concern.

4. Industrial Base and Supply Chain – The scale of the SBSP reference designs would aggravate industrial base and supply chain hurdles, including raw material and skilled workforce availability, that are common to all space missions. Our work did not assess whether the aerospace industry can realistically support the level of effort required for SBSP and what changes, if any, would be required to support this demand.

5. Launch Cadence – Our study assumed SpaceX's advertised Starship launch cadence, but how realistic that is remains to be seen. Nor is it clear how many of those flights would be available to launch an SBSP system. Additional analysis should be done on the feasibility of SBSP under multiple launch-cadence scenarios, in part to determine the minimum cadence required to deploy a system in the desired timeframe. The analysis may also consider the realism of SpaceX's planned launch cadence.

6. Propulsion – EP is cleaner but slower than chemical propulsion. However, advances are being made in EP technologies that may reduce the timeline to assemble an SBSP system. We recommend monitoring these advances to determine whether they make EP a more feasible option for SBSP.

7. Efficiencies of Scale – Our analysis looked at the costs to get to the first full-scale system in place, but did not consider whether efficiencies would be realized over time as additional SBSP systems are deployed. Additional analysis should be conducted to assess whether there are opportunities to lower the costs of future SBSP systems once the initial capability is demonstrated.

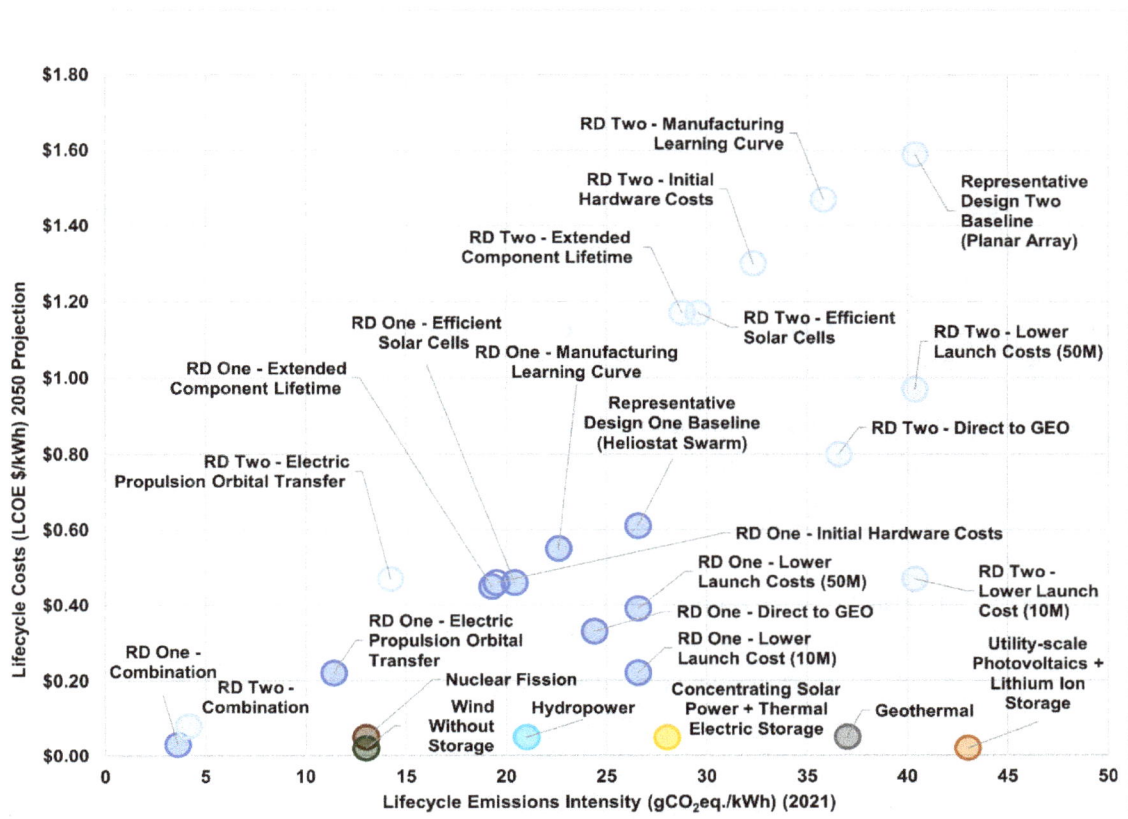

Figure 15. SBSP System Cost and GHG emissions with Sensitivity Analyses and Terrestrial Alternatives.

Appendix A: Representative Design Details

Two design references were assessed in this study: Representative Design One (RD1) or "Innovative Heliostat Swarm" and Representative Design Two (RD2) or "Mature Planar Array." Both designs are assembled and operate in GEO. The Aerospace Corporation assessed each of the design references based on source materials. NASA updated elements of these designs with more recent technical data, as seen in the tables below, and scaled the systems to each deliver about 2 GW of power to the grid. Both systems operate in geostationary orbit. Details on the concept of operations and attendant calculations are provided in Appendix B.

Innovative Heliostat Swarm is broadly derived from the Alpha Mark III concept (Mankins, SPS-Alpha Mark-III and an achievable roadmap to space solar power, 2021). This design uses reflectors and a concentrator to focus sunlight throughout each day.

Mature Planar Array is broadly derived from the Japan Aerospace Exploration Agency (JAXA) Tether and Caltech SSPP design (Sasaki, A new concept of solar power satellite: Tethered-SPS, 2006), (Sasaki, Demonstration Experiment for Tethered-Solar Power Satellite, 2009), (Pellegrino, 2022). This design uses flat panels, with solar cells facing away from the Earth and radiofrequency (RF) emitters facing toward the Earth. This design has a lower capacity factor due to limited capability to reposition itself or redirect sunlight toward its solar cells.

This study did not fully account for the changes in the Sun's position over the year, or how multiple RD2 systems in orbit may increase the capacity factor of the entire architecture. Multiple systems would increase the capacity factor; however, landmass on the Earth does not allow for even distribution across the day-night cycle, because much of the equator is underwater. This study is concerned with comparing representative examples of SBSP. Others may take the model developed for this study to further develop their own, more detailed analyses.

The Aerospace Corporation provided data and analysis in the form of interim reports, a methodology paper, and spreadsheet.

Table 4. System Specifications as Applied in this Study

Space-Based Segment				
Functional Step	**Parameter**	**RD1**	**RD2**	**Source**
	Distance to Earth surface (km)	35,786.00	35,786.00	Mankins, Sasaki
	Scaling factor	0.77	5	
	Solar panel size (km2)	11.473	19	Mankins, Sasaki
	Solar panel surface area (m2)	11,473,000.00	19,000,000.00	Calculated
Collect	Solar Constant (w/m2)	1,367.50	1,367.50	NASA
	Incident solar energy (MW)	15,689.33	25,982.50	Calculated
	Illumination	Secondary reflector array ~2-3 suns	No additional optical array 1 sun	Mankins, Sasaki
	Solar Cell Efficiency	0.35	0.35	Rodenbeck et al.
Convert in Space	DC to DC Conversion Efficiency	0.9	0.9	Rodenbeck et al.
	DC to RF Conversion Efficiency	0.7	0.7	Rodenbeck et al.
	Antenna Emission Efficiency	0.9	0.9	Rodenbeck et al.
	Atmospheric Travel Efficiency	0.98	0.98	Rodenbeck et al.
Transmit	Beam Collection Efficiency	0.95	0.95	Rodenbeck et al.
	RF frequency (GHz)	2.45	5.8	Mankins, Sasaki
Ground-Based Segment				
Functional Step	Parameter	RD1	RD2	Source
Receive	Rectenna Array Reception Efficiency	0.78	0.78	Rodenbeck et al.
	Diameter ground rectenna (km) receptor	6	4	Inferred from Mankins, Sasaki
Convert on Ground	DC to DC Conversion Efficiency	0.9	0.9	Rodenbeck et al.
Deliver	Capacity factor (% of year generating power)	0.997	0.6	Mankins, Sasaki
	Power delivered (MW)	2028.791344	2021.948467	Rodenbeck et al.

Appendix B: Methodology

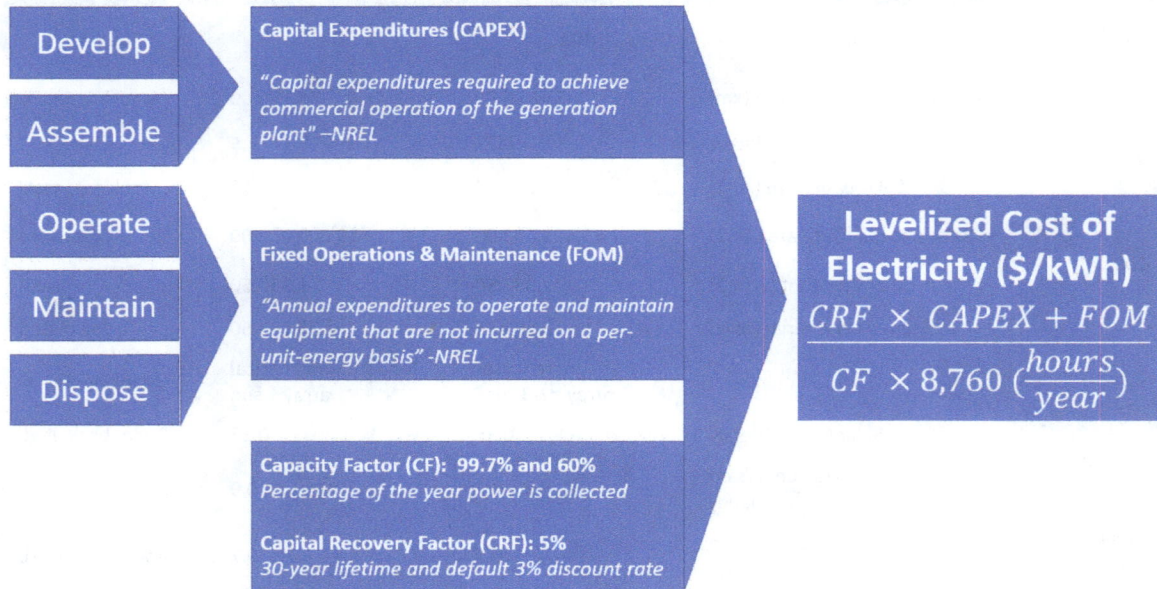

Figure 16. Cost Calculations for SBSP Systems

This appendix describes the approach to determining cost calculations for two representative space-based solar power designs—Representative Design One (RD1), the Innovative Heliostat Swarm, and Representative Design Two (RD2), the Mature Planar Array—and the methods of estimating their GHG emissions.

We describe the methodology in terms of the logical flow of each piece of the analysis. After a brief methodology overview, we proceed with sections as follows:

- Approach to Cost Calculations: This describes the functional decomposition of each system, details system costs by ConOps phase, and calculates the levelized cost of electricity (LCOE) for each system using determined capital expenditures and fixed operations and management costs. This also includes a breakdown of all relevant parameters and outputs, or results.
- Approach to GHG emissions Calculations: This includes an Economic Input Output Life Cycle Assessment to estimate emissions and a comparison to other sources of renewable energy technologies.
- Sensitivity Analyses: This describes alternate considerations that may have significant effects on lifecycle costs, including modified parameters, rationale, and effects on cost and GHG emissions.

Overview

The Aerospace Corporation was tasked by NASA's Office of Technology, Policy, and Strategy (OTPS) with conducting an Analysis of Alternatives to evaluate SBSP systems and inform NASA leadership on their costs and benefits as compared to other sustainable energy sources. Aerospace considered two systems — SPS-ALPHA Mark III and the Tethered Solar Powered Satellite — reviewed source documentation for each system, considered representative design reference systems for comparison, and decomposed each into modular components. To summarize each design reference system:

- The Innovative Heliostat Swarm is broadly derived from the SPS-ALPHA Mark III concept (Mankins, SPS-Alpha Mark-III and an achievable roadmap to space solar power, 2021). This design reference is a lower TRL modular design that can generate power for 99.7% of the year because of its concentrator and reflectors. The Mankins concept proposes that one system, with a mass of 7,500 MT, is capable of delivering 2 GW to one ground-based site.

- The Mature Planar Array is broadly derived from the JAXA Tethered Solar Power Satellite design: (Sasaki, A new concept of solar power satellite: Tethered-SPS, 2006), and updated with elements of Caltech's SSPP design: (Pellegrino, 2022). This is a relatively higher TRL concept consisting of a large panel and a bus system. One system can generate power for 60% of the year. The Sasaki concept proposes that each system, with a mass of 20,000 MT, is capable of delivering an average of 0.75 GW (and 1.2 GW maximum) to one ground-based site, while the 2023 in-space demonstration of Caltech's technology suggests a design with less mass is possible.

Aerospace assessed overall system specifications and made informed assumptions about information that was not available in the source material. Sources describe their systems as modular in nature; however, they have different module mass and size assumptions for each subsystem. To simplify calculations, Aerospace decomposed the SBSP systems into modular hardware units. The mass of one of these modular hardware units is derived by the ratio of each SBSP subsystem's mass to the total SBSP system mass (power, structure, attitude and determination control, propulsion, telemetry tracking and command, command and data handling, and thermal). Other spacecraft hardware assessed was not divided into modular units, and include servicer spacecraft and active debris removal spacecraft. Aerospace assessed the research, development, deployment, operation, and disposal of SBSP systems based on NASA cost modeling work breakdown structure (WBS) elements. Aerospace then used a combination of proprietary models and NASA models to calculate fiscal costs in a combination of statistical and parametric methods. Final costs were provided according to the following WBS: space hardware, land use, ground receiver of SBSP systems, access to space, orbital assembly, operations of the space segment, maintenance of the space segment, and debris mitigation.

These inputs were taken by NASAOTPS and validated with another model using 87 parameters (not including separate calculations for manufacturing learning curves, first order delta v, or assembly schedule). NASA OTPS applied specific decompositions of SBSP system functions and ConOps phases. The six functions are: collect, convert, transmit, receive, convert, and deliver. The five ConOps phases are: develop, assemble, operate, maintain, and dispose. NASA OTPS further modified, decomposed, and reorganized Aerospace data into these six functions and five phases, representing 87 parameters. The functions provided key inputs to electricity generation and delivery, while the phases provided key inputs to deriving the cost of SBSP systems.

To assess the SBSP systems' GHG emissions, NASA OTPS performed an Economic Input Output Life Cycle Assessment (EIO-LCA). This method is based on aggregate sector-level economic data. By adding environmental impact data to quantified direct and indirect economic inputs of purchases, environmental implications for select economic activity is derived. We use a mix of mass- and spend-based assessments, preferring mass wherever possible. The International Aerospace Environmental Group compiles EIO-LCA data from various sources (Carnegie Melon, DoD, and more). A key limitation of the EIO-LCA method is the assumed relationship between cost, efficiency, and GHG emissions; wherever possible, we used mass-based measures.

Aerospace provided data and analysis, including rationales for their methodologies, to NASA OTPS. To address the two questions in this study, we (OTPS) leveraged the data provided by Aerospace and performed an analysis to independently validate ConOps elements (such as orbital transfer and assembly times) and cost outputs for each design reference system and to estimate the GHG emissions of developing and operating each system. We also gathered data from authoritative sources on other electricity production technologies for comparison with the two design reference systems, primarily from NREL. We conclude with sensitivity analyses to explore potential effects on SBSP system costs of modifying parameters which could reasonably vary. The following sections detail our approach to cost calculations, GHG emissions, and methods to support sensitivity analyses. See Table 5 for a summary of our approach to the calculations.

Table 5. Analysis and Calculations Flow

Decomposition and Element Fiscal Costs	ConOps and Lifecycle Cost Calculation	Lifetime Costs to Generating Electricity	Climate Impact Calculations
SBSP system has 6 functions: collect, convert, transmit, receive, convert, deliver. Aerospace decomposed the collection of subsystems that perform these six functions, then calculated fiscal costs according to a WBS structure. NASA OTPS validated the findings with their own model following the functional decomposition into 89 parameters.	NASA OTPS arranged the 87 parameters into a ConOps to calculate the lifecycle cost to develop, assemble, operate, maintain, and dispose of the SBSP system.	NASA OTPS calculated the lifetime costs to generate electricity using the cost of each ConOps phase. This data was used to estimate LCOEs for the baseline and sensitivity analyses.	NASA OTPS performed an Economic Input Output Life Cycle Assessment using a mix of mass- and spend-based assessments. This data was used to estimate CO_2 equivalent per kWh for the baseline and sensitivity analyses.

Approach to Cost Calculations

Functional Decomposition of SBSP Systems

SBSP is assessed as having six functional steps, denoted by applicability to either in-space or ground-based segments.

- Functional steps for the in-space segment encompass each system's ability to: collect solar energy in space, convert solar energy to microwave radiation, and transmit microwave radiation to the Earth.
- Functional steps for the ground-based segment refer to each system's capacity to: receive microwave radiation at ground station rectenna, convert microwave radiation to electricity, and deliver electricity to the grid for use.

To describe each functional step, we use a combination of 1) the 89 parameters (including WBS elements) developed by NASA OTPS and the Aerospace Corporation, 2) design reference parameters provided by or inferred from Mankins (Mankins, SPS-Alpha Mark-III and an achievable roadmap to space solar power, 2021), Sasaki (Sasaki, A new concept of solar power satellite: Tethered-SPS, 2006), and Pellegrino (Pellegrino, 2022), 3) efficiency losses provided by an authoritative public-private study (Rodenbeck, et al., 2020), and 4) independently verified calculations determined by the study team. The following summarizes determinations for all parameters in the six functional steps. For a breakdown of each parameter by its corresponding functional step, along with values for both design reference systems, please refer to Tables 6 and 7 below.

From original panel sizes provided by Mankins and Sasaki, we applied the solar constant, as well as authoritative assessments of efficiency losses at each step, estimated by (Rodenbeck, et al., 2020). We normalized the outputs to 2 GW, which provided a "scaling factor" for each system. We evaluated each functional step through the lens of each system's scaling factor—0.77 and 5 applied to the Innovative Heliostat Swarm and Planar Array, respectively. We based system design

40

parameters off of Mankins and Sasaki (e.g., distance to Earth, solar panel size, illumination, radiofrequency, and capacity factor). The practice of scaling total system mass to solar cell efficiency comes from earlier literature (Mankins, SPS-ALPHA: The First Practical Solar Power Satellite via Arbitrarily Large Phased Array, 2012). Based on the scaling factor and solar panel size from Mankins and Sasaki, we calculated the total solar panel surface area. We then applied the solar constant (NASA, 2023) to determine the incident solar energy. Based on the power delivered, distance, and beam frequency from Mankins and Sasaki, Aerospace inferred the diameter of the ground rectenna receptor for each system. Changes to the angle of incident solar energy over the year were not evaluated, nor were changes in capacity factor from having multiple systems in space for RD2.

The following is a summary of major losses of efficiency for each functional step:
- Collect: solar cell efficiency (35%)
- Convert in-space: solar energy to microwave radiation (DC-DC 90% and DC-RF 70%)
- Transmit: antenna emission (90%), atmospheric travel (98%), and beam collection (95%)
- Receive: rectenna array reception (78%)
- Convert on-ground: DC-DC (90%)

Taking the scaling factor for each system and inefficiencies into account, and incorporating each system's capacity factor, results in final power delivery of approximately 2 GW (or about 13% of the incident solar energy).

ConOps Phases

We further break each system into five ConOps phases to evaluate costs by each phase of the full operational lifecycle. Evaluating costs by ConOps phase also allows us to estimate capital expenditures (CapEx) and fixed operations and management (FOM) costs, which we use in the next section to compare the cost of different renewable energy technologies.

Table 6. Functional Decomposition of Design Reference Systems for the Space-Based Segment

Space-Based Segment				
Functional Step	Parameter	RD1	RD2	Source
Collect	Distance to Earth surface (km)	35,786.00	35,786.00	Mankins, Sasaki
	Scaling factor	0.77	5	Calculated
	Solar panel size (km2)	11.473	19	Mankins, Sasaki
	Solar panel surface area (m2)	11,473,000.00	19,000,000.00	Calculated
	Solar Constant (w/m2)	1,367.50	1,367.50	NASA
	Incident solar energy (MW)	15,689.33	25,982.50	Calculated

Space-Based Segment				
	Illumination	Secondary reflector array ~2-3 suns	No additional optical array 1 sun	Mankins, Sasaki
Convert in Space	Solar Cell Efficiency	0.35	0.35	Rodenbeck et al.
	DC to DC Conversion Efficiency	0.9	0.9	Rodenbeck et al.
	DC to RF Conversion Efficiency	0.7	0.7	Rodenbeck et al.
	Antenna Emission Efficiency	0.9	0.9	Rodenbeck et al.
Transmit	Atmospheric Travel Efficiency	0.98	0.98	Rodenbeck et al.
	Beam Collection Efficiency	0.95	0.95	Rodenbeck et al.
	RF frequency (GHz)	2.45	5.8	Mankins, Sasaki

For a descriptive summary of the full lifecycle of SBSP systems, a description of each ConOps phase is described in Table 7. Note that the Develop ConOps phase includes technology development and manufacturing of bus hardware for the Planar Array design reference system.

Table 7. Functional Decomposition of Design Reference Systems for the Ground-Based Segment

Ground-Based Segment				
Functional Step	Parameter	RD1	RD2	Source
Receive	Rectenna Array Reception Efficiency	0.78	0.78	Rodenbeck et al.
	Diameter ground rectenna (km) receptor	6	4	Inferred from Mankins, Sasaki
Convert on Ground	DC to DC Conversion Efficiency	0.9	0.9	Rodenbeck et al.
Deliver	Capacity factor (% of year generating power)	0.997	0.6	Mankins, Sasaki
	Power delivered (MW)	2028.791344	2021.948467	Rodenbeck et al.

Table 8. Description of Each ConOps Phase

ConOps Phase	Description
Develop	Research and develop technologies
	Manufacture SBSP module hardware
	Perform project management, systems engineering, and mission assurance
Assemble	Manufacture servicers
	Launch SBSP modules and servicers to LEO
	Refuel launchers in LEO for orbital transfer to GEO

ConOps Phase	Description
Operate	Assemble SBSP modules in GEO with servicers
	Perform mission operations and data analysis to assemble
	Construct ground facilities
	Perform mission operations and data analysis to operate during service lifetime
	Manufacture spacecraft shielding
Maintain	Manufacture replacement SBSP modules and servicers
	Launch replacement SBSP modules and servicers to LEO
	Refuel launchers in LEO for orbital transfer to GEO
	Assemble replacement SBSP modules with replacement servicers in GEO
	Perform mission operations and data analysis to maintain
Dispose	Manufacture active debris removal spacecraft
	Launch active debris removal spacecraft to LEO
	Refuel launchers for orbital transfer to GEO
	Transfer all SBSP modules from GEO to graveyard orbit with active debris removal spacecraft
	Perform mission operations and data analysis to dispose

The following is a breakdown of: 1) what each phase encompasses, 2) relevant parameters and WBS elements, and 3) how the methodology supports cost outputs.

ConOps Phase 1: Develop

Develop Components

- Research and develop technologies
- Manufacture SBSP module hardware
- Perform project management, systems engineering, and mission assurance (PM/SE/MA)

The following is a list of all WBS Elements supporting this phase, including how they compare to both design reference systems. Note that WBS Elements map to either the components of this phase or to the specifications which serve as inputs to the components.

Table 9. Summary of all WBS Elements in the Develop ConOps Phase

WBS Elements	RD1	RD2	Sources & Notes
Thermal management	passive	passive	Aerospace
Power Mass per Module (kg)	2.00	2.50	Aerospace
Structure Mass per Module (kg)	1.00	1.25	Aerospace

WBS Elements	RD1	RD2	Sources & Notes
Attitude Determination and Control System Mass per Module (kg)	0.20	0.25	Aerospace
Propulsion Mass per Module (kg)	0.00	0.00	Aerospace
Telemetry Tracking and Command Mass per Module (kg)	0.20	0.25	Aerospace
Command and Data Handling Mass per Module (kg)	0.20	0.25	Aerospace
Thermal Mass per Module (kg)	0.40	0.50	Aerospace
Module Mass (kg)	4.00	5.00	Aerospace
Estimated number of modules	1,463,000.00	2,000,000.00	Inferred from Mankins, Sasaki, Pellegrino
SBSP In-space Mass (kg)	5,852,000.00	10,000,000.00	Inferred from Mankins, Sasaki, Pellegrino
Complexity/risk	High 2 elements reflector + PV-RF; Assembly in space	Low 1 element PV-RF sandwich	Aerospace inferred from Mankins, Sasaki
Technology Development Cost	$2.4B	$2.07B	Assuming TRL 4 (35%) for RD1 and TRL 6 (25%) for RD2. NASA OTPS modified, Aerospace inferred from Mankins, Sasaki with Design Maturity Factor and (Brunner & Jack, 2006)
PV-RF module	Modular sandwich	Modular sandwich (no wired signal/power interfaces required between the modules)	Mankins, Sasaki
Illumination	Secondary reflector array ~2-3 suns	No additional optical array 1 sun	Mankins, Sasaki
PV array	1 sided	bifacial	Mankins, Sasaki
Module Initial Cost	$1,000,000.00	$1,000,000.00	Analogs of satellite and PV industry
Module Manufacturing Learning Rate	0.75	0.75	Aerospace 11/2022
Module Manufacturing Cost ($)	$6.9B	$8.3B	Crawford model learning curve applied to initial cost and number of units.
Develop Program Management, Systems Engineering, and Mission Assurance Cost ($)	$2.3B	$2.6B	Aerospace Corporation: Mission PM/SE/MA costs are assumed at 25% of combined Hardware and Technology Development.
Total Develop Cost ($)	$11.6B	$12.9B	

The above data was used to derive the total hardware manufacturing, technology development, and mission management costs incurred for each design reference.

Estimating costs for manufacturing SBSP module hardware: Aerospace derived mass by taking the total system mass and averaging out components by the number of modules. This "bottoms-up" approach includes the following hardware subsystems: power, structure, attitude and determination control, propulsion, telemetry tracking and command, command and data handling, and thermal. As previously mentioned, Aerospace used a combination of cost models for hardware costs due to the maturity of both design references and the timeframe for implementation. (Note that this is an idealized decomposition of module subsystems for a comparative analysis. Mankins proposes separate modules for different functions, including assembly. Sasaki only modularizes the array for efficient terrestrial manufacturing.)

We referenced Mankins and Sasaki for number of modules in one system and applied the scaling factor to calculate the number of modules needed for 2 GW of output. Using the module mass and the number of modules needed, we estimated the total in-space mass, accounting for technology advances since Sasaki's original 2006 concept paper for the Planar Array design reference with Pellegrino et al.'s more recent technology demonstration.

Crawford model learning curve formula: cumulative cost of production = first unit cost * (learning rate * (natural log(total number of units produced) / natural log(2)))

We used Aerospace's learning curve estimate of 75% to account for efficiencies over time due to volume, arriving at final module development costs. We use the Crawford model learning curve formula instead of the Cumulative Average approach used by Aerospace. There are many other learning curve models, such as Stanford-B, Plateau, and Dejong's, and the study team encourages those seeking to repeat this analysis to experiment with alternatives. We assume an initial module cost of $1M. To arrive at final module development costs with our updated number of modules (taking the scaling factor into account), we developed a python script to apply the learning curve and deliver the cost of producing each individual module, a running average cost of all modules, and a running cumulative cost of all modules. The module manufacturing cost presented here is the cumulative cost of all modules. The following code was used to develop these estimates:

```
import numpy as np
import pandas as pd
import matplotlib.pyplot as plt

# Input variables
learning_rate = 0.75
first_unit_cost = 1000000
units_produced = list(range(1,6000001))
```

```
# Output variables
this_unit_cost = first_unit_cost * (learning_rate ** (np.log(units_produced) / np.log(2)))
cumulative_cost = np.cumsum(this_unit_cost)
average_cost = cumulative_cost / units_produced

# Graphs and tables
plt.plot(units_produced, cumulative_cost)
plt.xlabel('Units Produced')
plt.ylabel('Cumulative Cost')
plt.show()

df = pd.DataFrame({'Units Produced': units_produced, 'This Unit Cost': this_unit_cost,
'Cumulative Cost': cumulative_cost, 'Average Cost': average_cost})
print(df)

# Export to CSV
df.to_csv('C:/Users/use/folder/sbsp manufacturing learning curve.csv', index=False)
```

Estimating costs for technology development: Aerospace inferred technology development costs from successful NASA Science Mission Directorate missions, Mankins and Sasaki, and Brunner et al. (Brunner & Jack, 2006). Aerospace applied the same cost factor to both systems but given the recent technology demonstration by Caltech (Caltech, 2023), we adjusted RD2's TRL level to 6. Technology development costs are estimated to be 35% of hardware costs for RD1 (assuming TRL 4), and 25% of hardware costs for the RD2 (assuming TRL 6).

Estimating costs for mission project management, systems engineering, and mission assurance: Aerospace estimated in-space mission operations costs using a proprietary tool, the Mission Operations Cost Estimation Tool (MOCET), and a combination of NASA Science and commercial constellation mission analogies. Mission PM/SE/MA costs are assumed at 25% of combined spacecraft Hardware, Shielding, Maintenance Hardware, and Technology Development.

ConOps Phase 2: Assemble

Assemble Components
- Manufacture servicers
- Launch SBSP modules and servicers to LEO
- Refuel launchers in LEO for orbital transfer to GEO
- Assemble SBSP modules in GEO with servicers
- Perform mission operations and data analysis to assemble
- Perform mission support services of program management, systems engineering, and mission assurance

The following is a list of all WBS Elements supporting this phase, including how they compare to both design reference systems. Note that WBS Elements map to either the components of this phase or to the specifications which serve as inputs to the components.

Table 10. Summary of all WBS Elements in the Assemble ConOps Phase

WBS Elements	RD1	RD2	Sources
Assembly time per module (hours)	0.67	0.63	Aerospace, citing ISS and Orbital Express as analogs. Given launch vehicle fairing limitations, the Planar Array cannot be unfurled and must be assembled on orbit.
Modules assembled per servicer per month	1095	1152.631579	How many modules a single servicer can assemble in one month of continuous operation.
Total assembly time (work months, 1 servicer)	1336.073059	1735.159817	How many months it would take a single servicer to assemble the entire system under continuous operation.
Servicer mass (kg)	2,800.00	2,800.00	https://nssdc.gsfc.nasa.gov/nmc/spacecraft/display.action?id=2020-056B
First Servicer Cost ($)	$1,000,000,000.00	$1,000,000,000.00	Aerospace, citing (Messier, 2020).
Number of servicers	17	15	Aerospace, Mankins, Sasaki, NASA OTPS amended.
Servicer Manufacturing Learning Rate	0.85	0.85	Aerospace assessment of analogs.
Time to assemble SBSP system (months)	89	152	Simple model given launch cadence, transit times, payload capacity, minimizing servicer downtime.
Maximum Modules per Starship	25000	20000	Starship payload capacity / module mass
Module: Servicer total ratio	86,058.82	133,333.33	Total servicers to modules (number of servicers determined using assembly model to minimize servicer downtime)
Assembly time of averaged ratio of modules to servicers (years)	6.55	9.64	
Number of Servicers per Payload	0.288155129	0.149372635	How many servicers there are in a single payload, average. Assemble ConOps tab assumes earlier payloads carry more servicers.
Time to refuel (months)	1	1	Estimate, does not consider cryogenic boiloff in orbit.
Orbital transfer time (months)	1	1	Output of first order delta v calculation

WBS Elements	RD1	RD2	Sources
Servicer Manufacturing Cost - Assemble ($)	$10.9B	$9.8B	Using the Crawford model, 85% learning curve.
Assemble Program Management, Systems Engineering, and Mission Assurance Cost ($)	$2.7B	$2.5B	25% of hardware cost.
Assemble Mission Operations cost ($/month)	$1,200,000.00	$1,200,000.00	4x Aerospace's ADR fleet 300k monthly operations cost. Analog of Starlink (300 seats 200k each estimate), annual operations cost https://www.reuters.com/article/us-spacex-starlink-insight-idUSKCN1N50FC
Total Mission Operations Assembly cost not including launch ($)	$106M	$182M	Aerospace, modified proportional to number of servicers; MOCET historical average of NASA missions used. Aerospace ADR fleet was assumed autonomous and had flat operations cost; we are applying a similar methodology evenly across each phase's mission operations costs.
Starship payload capacity to LEO (kg)	100,000.00	100,000.00	https://www.spacex.com/media/starship_users_guide_v1.pdf
Starship launch cost ($)	$100,000,000.00	$100,000,000.00	Aerospace 9/2022 cites (Vorbach, 2022)
Block buy discounted rate	0.85	0.85	Aerospace
Disposable Starship payload capacity to GEO (kg)	21,000.00	21,000.00	For sensitivity analysis https://www.spacex.com/media/starship_users_guide_v1.pdf
Refueled Starship payload capacity to GEO (kg)	100,000.00	100,000.00	https://www.spacex.com/media/starship_users_guide_v1.pdf
Refuel launches required per Starship LEO to GEO	12.00	12.00	*Blue Origins Fed'n, LLC; Dynetics, Inc.-A Leidos Co., supra* at 27 n.13.
Reuses of each Starship	100.00	100.00	BryceTech SME for optimistic value, SpaceX
Total Develop Upmass (kg)	5,899,600.00	10,042,000.00	
Disposable launches total	281	479	
Total Starship Assemble hardware launches	59	101	How many Starships are carrying hardware to GEO in this phase.
Refuel launches Assemble total	708	1212	How many additional Starship launches needed to refuel all hardware-carrying Starships to GEO in this phase.
Total Launch Costs Assemble ($)	$65.2B	$111.6B	

WBS Elements	RD1	RD2	Sources
Number of Starships built (reusable)	66	113	Assumes Starships to GEO are never reused.
Starship Capable launch sites capacity (launches per annum)	104	104	At present, Starship-capable launch sites: Boca Chica (~5/year), LC39A (<20/year), new LC40 pad (unknown capability), and an unknown number of off-shore platforms - we assume capability to launch 2 Starships every week (average over period of time under study). BryceTech SME.
Hardware payloads to GEO per annum	8	8	
Time to launch all payloads (years)	7.375	12.625	Based on first order assembly model
Total Assemble Cost ($)	$78.9B	$124.1B	

The above WBS elements were used to derive the cost to launch original hardware, including refuel launches, servicer manufacturing costs, and mission operations and data analysis costs for each design reference.

Estimating costs for Launch of Original Hardware: Aerospace approached launch costs by estimating the total launches needed to deliver the SBSP systems to their operational orbit, which includes launches to refuel in LEO. Aerospace conducted a trade study to find the most desirable launch vehicle from a mass- and volume-to-orbit standpoint. After evaluating multiple heavy-lift launch and transfer vehicles and also considering ability to refuel and reuse, Starship was selected.

Launch of original hardware includes both launch costs for initial assembly and launch costs to refuel in LEO. To find the number of launches for initial assembly, we divide the total upmass (the total in-space mass, normalized to 2 GW, plus the total servicer mass) by Starship's payload capacity to LEO. Starship's capacity to LEO is based on the 2020 users guide published by SpaceX. To find the number of launches to refuel in LEO, we assume that for every payload of original hardware delivered to LEO, it will require 12 additional Starship launches to refuel for transfer to GEO (Blue Origin Fed'n, LLC; Dynetics, Inc.-A Leidos Co., 2021). After finding the total number of launches for original hardware and additional fuel, we multiply by Starship's launch cost and also apply a block discount rate of 15%. Starship's launch cost and the block discount rate were both determined by Aerospace.

Estimating costs for Servicer Manufacturing: Costs for spacecraft servicer manufacturing were based on the Mission Extension Vehicle 2 (MEV-2). Aerospace assumed an initial cost of $1B for the first servicer unit. Based on a one-year design life for servicers, they estimated a 7-year assembly time, with an 85% learning curve. A Crawford Model learning curve formula was applied to estimate costs for all servicers.

Estimating assembly time: We assume a launch capacity of two per week or 104 per year. At this rate, it takes just over 6 weeks for 13 Starships to reach orbit. We therefore assume it takes 7 to 8 weeks for one payload-laden Starship to be fully refueled in LEO. We assume that no cryogenic boil off occurs in orbit. Given the variable possible conditions of launch sites and orbits, we assume that one additional month of travel time to GEO is more than sufficient. We assume 4,200 m/s is more than sufficient for a chemical propulsion orbital transfer from LEO to GEO. Applying known inputs to delta v and transit time equations (below) yields delta v well in excess of this minimum. For Starship, we use 355 s, 14.7 MN, a payload mass of 100 MT, fuel mass of 120 MT, and dry mass of 100 MT.

Delta V = specific impulse * standard acceleration due to gravity * natural logarithm (initial mass/final mass)

Transit time = (initial mass / thrust) * deltaV / specific impulse

Number of modules in GEO = (this month's number of payloads in GEO - previous month's number of payloads in GEO) * maximum number of modules on Starship - ((number of servicers in GEO this month - number of servicers in GEO last month) * servicer mass) / module mass + number of modules in GEO last month

Modules assembled per month = number of servicers in orbit * servicer monthly assembly rate

Finally, the cumulative number of modules assembled was evaluated at each month to maximize the efficiency of the servicer fleet by minimizing servicer idle time. We use Aerospace's assessment that servicers will take approximately 40 and 38 minutes to assemble a single RD1 and RD2 module, respectively. Our assembly timeline model seeks to minimize servicer downtime during construction phases. Our first order model estimates 7.4 and 12.6 years to assemble RD1 and RD2, with 17 and 15 servicers, respectively. In both cases, no servicer is idle until the final year of assembly operations.

Estimating costs for Mission Operations & Data Analysis – Assembly: Aerospace assessed the operation of the ADR fleet as autonomous, common to any fleet size, and at a rate of $300k/month. For the more complex task of assembling the SBSP system itself, we assume four times this cost, or $1.2M/month. This number is informed by analogs like Starlink (Johnson & Roulette, 2018), assuming one sixth of the Starlink workforce (50/300), assuming capital costs for labor ($18,000/month) and adding the result ($900k/month) to Aerospace assessed cost of operating an autonomous fleet.

ConOps Phase 3: Operate

Operate Components:
- Construct ground facilities
- Perform mission operations and data analysis to operate during service lifetime

- Develop spacecraft shielding
- Perform mission support services of program management, systems engineering, and mission assurance

The following is a list of all WBS Elements supporting this phase, including how they compare to both design reference systems. Note that WBS Elements map to either the components of this phase or to the specifications which serve as inputs to the components.

Table 11. Summary of all WBS Elements in the Operate ConOps Phase

WBS Elements	RD1	RD2	Sources
RF frequency (GHz)	2.45	5.8	Aerospace
Diameter ground rectenna (km) receptor	6	4	Aerospace derived from Mankins, Sasaki.
Operational lifetime (years)	30.00	30.00	
Capacity factor (% of year generating power)	0.997	0.6	Mankins, Sasaki
Hours in a year	8760	8760	
Lifetime kWh	533,166,365,206.45	531,368,057,222.62	
Operations cost ($/year)	14,400,000.00	14,400,000.00	
Mission Operations & Data Analysis - Operations Cost ($)	$432,000,000.00	$432,000,000.00	Aerospace, NASA OTPS derived
Ground System cost ($)	$3.1B	$8,27B	Aerospace derived from solar farms and very large arrays. We assume RD2 requires 5 rectennas.
PM/SE/MA of spacecraft shielding ($)	$49,2M	$133.25B	
Spacecraft Shielding cost ($)	$197,000,000.00	$533,000,000.00	Aerospace
Total cost of Operations ($)	$3.8B	$9.36B	

The above WBS elements were used to derive the cost of mission operations and data analysis to operate during the service lifetime, as well as ground system costs and spacecraft shielding costs for each design reference system.

Estimating costs for Mission Operations and Data Analysis – Operations: The above assumption of $1.2M/month was applied over the course of the SBSP systems' assumed 30 years of operation.

Estimating costs for Ground System: Aerospace derived costs for ground systems based on analogies of U.S.-based solar array farms. They evaluated 12 solar farms for cost estimates based on land use and ground-based large antenna arrays. We applied these results to the Innovative Heliostat Swarm without modification, and applied the scaling factor for the Mature Planar Array, which requires five receivers.

Estimating costs for Spacecraft Shielding: Aerospace assessed spacecraft shielding costs as part of the broader active debris removal assessment, which also includes remediation. We assess spacecraft shielding in the Operations ConOps phase since it is critical to ensuring operations and consider remediation as part of the Dispose ConOps phase.

The cost due to spacecraft shielding is the cost of the additional mass needed to protect from orbital debris. Aerospace used the same mass increase for both systems (0.74%), supported by the Aerospace-developed assessment, "Space Debris: Models and Risk Analysis."

ConOps Phase 4: Maintain

Maintain Components:

- Manufacture replacement SBSP modules and servicers
  - Assemble replacement SBSP modules with replacement servicers in GEO
- Launch replacement SBSP modules and servicers to LEO
  - Refuel launchers in LEO for orbital transfer to GEO
- Perform mission operations and data analysis to maintain
- Perform mission support services of program management, systems engineering, and mission assurance

The following is a list of all WBS Elements supporting this phase, including how they compare to both design reference systems. Note that WBS Elements map to either the components of this phase or to the specifications which serve as inputs to the components.

Table 12. Summary of all WBS Elements in the Maintain ConOps Phase

WBS Elements	RD1	RD2	Sources
All hardware lifetime (years)	10.00	10.00	Aerospace
Number of refurbishment cycles	2	2	Aerospace
Number of refurbishment modules	2,926,000.00	4,000,000.00	Mankins, Sasaki
Number of maintenance servicers	34	30	Mankins, Sasaki
Total Maintain Upmass (kg)	11,799,200.00	20,084,000.00	Sasaki
Disposable launches total	562	957	
Total Starship Maintain launches	118	201	
Refuel launches Maintain total	1416	2412	
Number of Starships built (reusable)	131	224	Assumes Starships to GEO are not reused
Cost of Maintenance Servicers ($)	$15B	$13.6B	
Cost of Maintenance Modules ($)	$6.2B	$7.47B	

WBS Elements	RD1	RD2	Sources
Maintain Program Management, Systems Engineering, and Mission Assurance Cost ($)	$5.3B	$5.27B	
Cost of Maintenance Launches ($)	$130.4B	$222.1B	
Cost of Maintenance Operations ($)	$213.6M	$364.8M	
Total Cost Maintain ($)	$157.15B	$248.85B	

The above WBS elements were used to derive cost estimates for all maintenance hardware, including launch and orbital assembly.

Estimating costs for Maintenance Hardware: Maintenance hardware includes the cost of maintenance modules, including replacements for modules and servicers. Assuming modules have a 10-year lifetime, thus requiring two refurbishment cycles in a 30-year period, and referencing Mankins and Sasaki for the number of maintenance modules, we arrive at the total number of maintenance modules.

We extend our prior application of the SBSP module and servicer spacecraft learning curves to determine the cost of two refurbishment cycles for all modules and servicers, respectively. Note that for all learning rates, Aerospace uses a model based on cumulative averages. We use the Crawford Model, which assumes slower rates of learning and results in higher total costs.

Estimating costs for Launch of Maintenance Hardware: Launch of maintenance hardware includes cost, using Starship, for all maintenance launches. Maintenance launches include launch of maintenance servicers and modules. Based on the upmass for the total number of maintenance modules, we find the total number of Starship launches needed to launch to LEO. We assume 12 refueling launches to transfer to GEO are needed for each launch and apply a 15% block buy discount rate (as in the Assemble phase).

Estimating costs for Orbital Assembly of Maintenance Hardware: We use the same methodology as we did for the initial assembly of the SBSP system, assuming of $1.2M/month for operations, and the same number of servicers, launches, and time to assemble for each refurbishment cycle.

ConOps Phase 5: Dispose

Dispose Components:
- Manufacture active debris removal spacecraft
- Launch active debris removal spacecraft to LEO
- Refuel launchers for orbital transfer to GEO
- Transfer all SBSP modules from GEO to graveyard orbit with active debris removal spacecraft
- Perform mission operations and data analysis to dispose

53

- Perform mission support services of program management, systems engineering, and mission assurance

The following is a list of all WBS Elements supporting this phase, including how they compare to both design reference systems. Note that WBS Elements map to either the components of this phase or to the specifications which serve as inputs to the components.

Table 13. Summary of all WBS Elements in the Dispose ConOps Phase

WBS Elements	RD1	RD2	Sources
Active Debris Removal Vehicle Base Cost ($)	$500,000,000.00	$500,000,000.00	Aerospace 11/2022
Active Debris Removal Operations Costs ($/month)	$300,000.00	$300,000.00	Aerospace 11/2022
ADR operations duration (months)	180.00	180.00	Aerospace
ADR Vehicle Mass (kg)	2,500.00	2,500.00	Aerospace
ADR Vehicle Payload Capacity (kg)	5,000.00	5,000.00	Aerospace
Active Debris Removal Vehicles Needed	59.00	100.00	Aerospace 11/2022
Active Debris Removal Management Costs ($)	$2.77B	$4.3B	Aerospace 11/2022
Active Debris Removal Hardware Costs ($)	$18.5B	$29B	Aerospace 11/2022
Active Debris Removal Ground Costs ($)	$1.85B	$2.9B	Aerospace 11/2022
Active Debris Removal Operations Costs ($)	$54,000,000.00	$54,000,000.00	Aerospace 11/2022
Active Debris Removal Launch Costs ($)	$1.5B	$2.55B	
Active Debris Removal Reserves ($)	Removed	Removed	Aerospace 11/2022
Total Active Debris Removal Costs ($)	$24.7B	$38.9B	Aerospace 11/2022

The above WBS elements were used to derive cost estimates for ADR vehicle manufacturing, management, operations, and ground systems costs, as well as associated launches.

Estimating costs for ADR Vehicle Manufacturing: Aerospace assessed an approach to ADR with ADR transfer vehicles acting as space tugs to move modules from GEO to a graveyard orbit. ADR transfer vehicles are based on Northrop Grumman's Mission Extension Vehicle (MEV). They are assumed to deliver modules to a graveyard orbit at the end of their design life (every 10 years), or twice during the mission lifecycle, and again at the 30-year mark for end of mission disposal. Aerospace did not consider launch costs for ADR transfer vehicles in the ADR methodology. For cost estimates, Aerospace considers mass estimates using the MEV in addition to in-house costing methodologies.

For our cost estimates, we leverage the same first-unit cost assessed by Aerospace ($500M/unit) and also apply the same 90% learning rate to account for improvement over time with experience in production. We determine the number of transfer vehicles needed based on the updated in-space mass using the scaling factor, the ADR transfer vehicle payload capacity, and duration of ADR operations.

Estimating costs for ADR Vehicle Management, Operations, and Ground Costs: Aerospace evaluated ADR management costs by applying a 15% multiplier to the cost of ADR hardware. We leverage this approach using updated ADR hardware costs taking the scaling factor into consideration.

ADR operations costs are based off of the operational cost for removal, estimated monthly, and the total duration of ADR operations. Aerospace provides an estimate of the monthly cost, assuming the same cost regardless of fleet size due to the use of autonomous operations. Duration of operations assumes that the refurbishment cycles and the end of mission cycle take five years each, totaling 15 years of operations.

Estimating costs for Launch Vehicle – Active Debris Removal: We approach launch costs similarly to Aerospace, using our normalized values and several updated parameters. Launch vehicle costs for ADR vehicles are determined based off the updated number of vehicles needed, the servicer mass (which is based off of MEV-2), Starship's payload capacity to LEO (estimated using the 2020 Starship Users Guide), and the number of refuel launches to transfer from LEO to GEO. We also apply a 15% block buy discount rate to the cost to launch Starship.

The following is a detailed list of all ConOps parameters supporting this phase, including a comparison of both design reference systems.

Determining Capital Expenditures and Fixed Operations and Management

Evaluating costs by ConOps phase also allows us to estimate CapEx and FOM costs. We follow NREL's (NREL, 2022) definition of CapEx, "Capital expenditures required to achieve commercial operation of the generation plant," and FOM, "Annual expenditures to operate and maintain equipment that are not incurred on a per-unit-energy basis."

Following these definitions, we consider the Develop and Assemble ConOps phases as contributors to CapEx costs. This accounts primarily for development and assembly of the in-space hardware, including launch. We group all other ConOps phases under FOM, as they are not required to initiate the first system.

CapEx and FOM are key inputs to determining the levelized cost of electricity (LCOE), which is a useful method of comparing the costs of different renewable energy technologies. We discuss this method in the following section.

Levelized Cost of Electricity

Overview

In addition to cost estimates evaluated over the lifecycle, we used fiscal cost estimates to perform a traditional energy cost analysis. This allows for a comparison of each system's costs per megawatt hour and an additional comparison to other renewable energy technologies, if produced from a similar power plant. As described by NREL (NREL, 2023), "Capital expenditures [are] required to

55

achieve commercial operation of the generation plant" and fixed operations and maintenance (FOM) are "[a]nnual expenditures to operate and maintain equipment that are not incurred on a per-unit-energy basis." To restate from the previous section, CapEx costs include the Develop and Assemble ConOps phases and FOM costs include the Operate, Maintain, and Dispose ConOps phases.

NREL is the most authoritative USG source for electricity production technology data. NREL's Annual Technology Baseline (ATB) (NREL, 2022) provides assessments of a range of technologies every year. We use the data published in 2022 and apply their Levelized Cost of Electricity (LCOE) formula to compare electricity production technology costs as uniformly as possible. As described by NREL, "The levelized cost of energy (LCOE) calculator provides a simple way to calculate a metric that encompasses capital costs, operations and maintenance (O&M), performance, and fuel costs of renewable energy technologies."

Our primary divergences from NREL's calculator are in the consideration of more refined financing inputs. We do not consider the rate on return of equity in our calculations for example, as our baseline scenario assumes Government financing. Different activities have different financing assumptions in the ATB (construction has a different financing rate for example). We apply the same financing rate to each design reference system and all comparison energy technologies. We also assume manufacturing of all activities occur in the U.S. and do not consider cost factors for different locations of work. Finally, we re-state NREL's assertion that this is a simplified calculation for comparative purposes and that a more detailed analysis to include financing, discounts, and other costs would provide a more holistic look.

Calculation

NREL calculates LCOE using the following formula:

LCOE = {(overnight capital cost * capital recovery factor + fixed O&M cost)/(8760 [hours in a year] * capacity factor)} + (fuel cost * heat rate) + variable O&M cost.

When applied to the two design reference systems, each variable is described as follows:
- Overnight Capital Cost: CapEx Cost / Power Delivered
- Capital Recovery Factor: Capital Recovery Factor,[†] with a default discount rate of 3%
- Fixed O&M Cost: FOM cost / Power Delivered / Lifetime
- Capacity Factor: Percentage of the year in which power is generated
- Fuel cost: N/A
- Heat rate: N/A
- Variable O&M cost: N/A[††]

†Capital Recovery Factor = $\{i(1 + i)^n\} / \{[(1 + i)^n]-1\}$ where "i" is the interest rate and "n" is the number of annuities received (n = 30, or the lifetime in years of each design reference system).

††Variable Operations and Maintenance covers electricity fuel sources, so NREL does not include this cost category for any renewable electricity production technology. SBSP does not use fuel as a direct input to electricity production.

In applying this calculation to comparative renewable energy technologies, all variables remain the same with exception of the removal of power output (~2 GW), stemming from the SBSP design reference systems' scaling factor. A breakdown of the inputs (all variables) is seen in Table 14.

Table 14. LCOE Calculations in detail

Renewable Energy Technology	Lifetime (Years)	Output (kW)	Capacity Factor (%)	Hours in a Year	CapEx ($/kW)	FOM ($/kW-year)	Variable O&M ($/kWh)	Fuel Cost ($/MMBtu)	Heat Rate (Btu/kWh)	CRF (default discount rate of 3%)	Levelized Cost of Utility Electricity 1 ($/kWh)	Levelized Cost of Utility Electricity 2 ($/kWh)	Simple Levelized Cost of Alternative Energy ($/kWh)	$/MWh Utility 1	$/MWh Utility 2	$/MWh	Notes
RD1 Baseline	30.00	2.03E+06	0.997	8760	$53,689.58	$4,823.97	$-	$-	0	0.051019259	$0.21	$0.26	$0.61	$210.00	$260.00	$610.07	
RD2 Baseline	30.00	2.02E+06	0.6	8760	$75,807.25	$4,680.37	$-	$-	0	0.051019259	$0.21	$0.26	$1.59	$214.00	$260.00	$1,590.13	
Land Wind 2050	30		0.47	8760	$760.00	$33.00	$-	$-	0	0.051019259	$0.21	$0.26	$0.02	$214.00	$260.00	$17.43	Class 4 wind
PV Utility 2050	30		0.29	8760	$618.00	$13.00	$-	$-	0	0.051019259	$0.21	$0.26	$0.02	$214.00	$260.00	$17.53	Class 5, 4 hrs Li-ion 50 MW capacity
PV Utility + Storage 2050	30		0.3	8760	$855.00	$19.00	$-	$-	0	0.051019259	$0.21	$0.26	$0.02	$214.00	$260.00	$23.83	Class 5
CSP + TES 2050	30		0.63	8760	$3,894.00	$56.10	$0.00	$-	0	0.051019259	$0.21	$0.26	$0.05	$214.00	$260.00	$49.06	Class 2
Nuclear 2050	60		0.94	8760	$5,892.00	$146.00	$0.00	$0.74	0.01044	0.036132959	$0.30	$0.37	$0.05	$302.00	$367.00	$54.31	AP 1000
Hydro 2050	100		0.34	8760	$2,471.00	$62.00	$-	$-	0	0.031646666	$0.45	$0.54	$0.05	$448.00	$544.00	$47.07	NPD 1
Geothermal 2050	30		0.9	8760	$5,080.00	$99.00	$-	$-	0	0.051019259	$0.21	$0.26	$0.05	$214.00	$260.00	$45.43	Hydro / Flash

Results

Total SBSP Lifecycle Costs

Table 15. Costs by Capital Expenditures and Fixed Operations and Management

	ConOps Phase	RD1 ($)	RD2 ($)
CapEx	Develop	11.65B	12.95B
	Assemble	78.9B	124.1B
	Total	$90.57B	$137B
FOM	Operate	3.79B	9.36B
	Maintain	157.1B	248.85B
	Dispose	24.7B	38.94B
	Total	$185.65B	$297.17B
Total		$276.2B	$434.2B

SBSP Lifecycle Costs by ConOps Phase

Table 16. SBSP Lifecycle Costs for the Develop ConOps Phase

Cost Estimate Results (FY22$)	RD1 Baseline ($)	RD2 Baseline ($)
Mission Program Management, Systems Engineering, and Mission Assurance	2.33B	2.59B
Technology Development	2.4B	2B
Spacecraft Hardware	6.9B	8.29B
Develop Total	11.65B	12.95B

Table 17. SBSP Lifecycle Costs for the Assemble ConOps Phase

Cost Estimate Results (FY22$)	RD1 Baseline ($)	RD2Baseline ($)
Launch of Original Hardware	65.19B	111.6B
Servicer Manufacturing	10.89B	9.86B
Mission Program Management, Systems Engineering, and Mission Assurance	2.7B	2.46B

Cost Estimate Results (FY22$)	RD1 Baseline ($)	RD2Baseline ($)
Mission Operations & Data Analysis - Assembly	106.8M	182.4M
Assemble Total	78.9B	124.1B

Table 18. SBSP Lifecycle Costs for the Operate ConOps Phase

Cost Estimate Results (FY22$)	RD1 Baseline ($)	RD2 Baseline ($)
Mission Operations & Data Analysis - Operations	432M	432M
Ground System	3.1B	8.27B
Spacecraft Shielding	246M	666M
Operate Total	3.79B	9.36B

Table 19. SBSP Lifecycle Costs for the Maintain ConOps Phase

Cost Estimate Results (FY22$)	RD1 Baseline ($)	RD2 Baseline ($)
Maintenance Modules and Servicers	21.2B	21.1B
Mission Program Management, Systems Engineering, and Mission Assurance	5.3B	5.27B
Launch of Maintenance Hardware	130.39B	222.1B
Orbital Assembly of Maintenance Hardware	213.6M	364.8M
Maintain Total	157.15B	248.85B

Table 20. SBSP Lifecycle Costs for the Dispose ConOps Phase

Cost Estimate Results (FY22$)	RD1 Baseline ($)	RD2 Baseline ($)
ADR Vehicle Manufacturing	18.5B	29B
ADR Vehicle Management, Operations, and Ground Costs	4.68B	7.3B

Cost Estimate Results (FY22$)	RD1 Baseline ($)	RD2 Baseline ($)
Launch Vehicle - Active Debris Removal	1.5B	2.55B
Dispose Total	24.7B	38.9B

Table 21. Total SBSP Lifecycle Costs

	RD1 Baseline ($)	RD2 Baseline ($)
Total $	$276B	$434B
Total $ / kg (space segment)	$47,203.68	$43,423.96
Total $ / kWh (lifetime)	$0.52	$0.82

Levelized Cost of Electricity

Table 22. LCOE Calculation results

Renewable Energy Technology	LCOE ($/kWh)	LCOE ($/MWh)
RD1 Baseline	0.61	610.07
RD2 Baseline	1.59	1,590.13
Land Wind 2050	0.02	17.43
Offshore Wind 2050	0.04	39.81
PV Utility 2050	0.02	17.53
PV Utility + Storage 2050	0.02	23.83
CSP + TES 2050	0.05	49.06
Nuclear 2050	0.05	54.31
Hydro 2050	0.05	47.07
Geothermal 2050	0.05	45.43

Approach to GHG Emissions Calculations

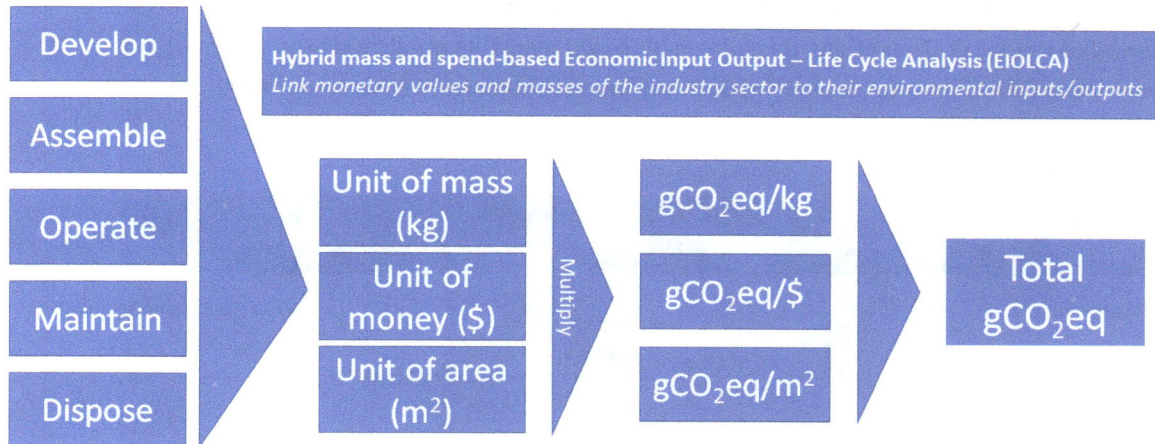

Figure 17. SBSP GHG emissions calculations graphical summary

GHG Emissions

$gCO_2eq./kWh = (\Sigma \ total \ kg * gCO_2eq./kg + total \ m2 * gCO_2eq./m2 + total \ \$ * gCO_2e.q/\$) / (system \ electricity \ generation * operational \ years * hours \ in \ a \ year)$

The Aerospace Corporation provided expert assessment of the bill of materials for modules, launch vehicles, servicers, and the ground segment. This gave us high-level estimates of how much steel, circuitry, or other material inputs were required for each hardware element, which in turn allowed us to estimate the total materials required.

However, several factors, such as transportation, processing, assembly, and more were not being accounted for. We spoke with an expert at the National Renewable Energy Laboratory (NREL) who suggested we consider the Economic Input Output Life Cycle Assessment (EIO-LCA) (Cobas-Flores, 1998) method. This method is based on aggregate sector-level economic data, such as the North American Industry Classification System (NAICS). Input-output analysis, developed by Wassily Leontief (Leontif, 1951) is a technique for capturing economy-wide interdependencies. For example, automobile manufacturing needs steel, and steel needs iron ore and coal, which in turn need automobiles to transport them from mines to factories. This method allows for a far more accurate understanding of changes to economic outputs of individual sectors as they ripple across the economy.

Data collected on environmental impacts by organizations like the Environmental Protection Agency (EPA) may also be organized by the same industry classification codes. By adding environmental impact data to quantified direct and indirect economic inputs of purchases, environmental implications for economic activities and their dependencies are derived. We use a mix of mass and spend-based assessments, preferring mass wherever possible. Carnegie Melon University and DoD

62

maintain databases with the gCO_2eq. of various mass, spend, and area NAICS categories. Unfortunately, the Carnegie Melon EIO-LCA website eiolca.org was nonfunctional during the entire period of this work. Thankfully, this information was available publicly via the International Aerospace Environmental Group (IAEG) (International Aerospace Environmental Group, 2023).

Table 23. Full list of parameters used for the GHG emissions assessment. Unless otherwise noted, source is IAEG, 2023.

Parameter	GHG emissions kqCO₂eq., Source
PV area in m2	0.27 / m2 (Liang & You, 2023)
Cost of modules	174 / kUSD, "Propulsion units and parts for space vehicles"
Total Prop Needed for Starship (kg) * Prop cost for Starship ($/kg) * Total Starship launches	1,290 /kUSD, Basic organic chemicals LOX is ~$0.16/kg and liquid methane ~$0.4/kg, ~3:1 ratio for starship fuel
Number of launches * launch vehicle emissions	MT 2,683 CO_2 + 1.7 NO (*298 for CO_2eq.)
Launcher metal mass (kg) * Total Starships manufactured	1.7 /kg Steel, unspecified process
Structure composite mass (kg) * Total Starships manufactured	495.06/kg Composite-based manufactured products
Harness Mass (kg) * Total Starships Manufactured	5.84/kg Copper unspecified process
{[Electronics mass (kg) + electronics structure mass (kg)]/0.181437}* 0.064516 * Total Starships Manufactured	606.12 /m2, printed circuit board; printed circuits 0.181437kg = 0.064516 m2 (Silver Circuits, n.d.)
Launcher metal mass (kg) * Total Starships manufactured	
Structure composite mass (kg) * Total servicers manufactured	
Harness Mass (kg) * Total servicers Manufactured	
{[Electronics mass (kg) + electronics structure mass (kg)]/0.181437}* 0.064516 * Total servicers Manufactured	
Spacecraft shielding (kg)	3.75 Glass fibers, high strength, unspecified process (kgCO₂eq./kg)
Rectenna cost ($)	182.89 /kUSD, Wireless communication equipment; assumed 20% of ground segment cost goes to rectenna

Parameter	GHG emissions kgCO$_2$eq., Source
R&D cost ($)	250 Research, development, and testing services (kgCO$_2$eq./kUSD)
Mission Program Management, Systems Engineering, and Mission Assurance cost ($)	163.11 Marketing, research and other miscellaneous professional, scientific, and technical support services (kgCO$_2$eq./kUSD)
Concrete building surface area (m2)	825 /m2 Industrial building, concrete and assuming similar sized buildings
Total rectenna support structure mass (kg)	Assumed steel
system electricity generation * operational years * hours in a year	~2,000,000 kWh * 30 * 8760 = 525,600,000,000 kWh

Table 24. Full list of GHG emissions calculations by ConOps phase

Key input parameters	RD1	RD2	kgCO$_2$eq.	Source
Develop				
PV area in m2	11,473,000.00	19,000,000.00	0.27	(Liang & You, 2023)
PV GHG emissions (kgCO$_2$eq.)	3,097,710.00	5,130,000.00		
Cost of modules minus PV	$3,453,065,585.02	$4,146,166,077.75		
GHG emissions of modules minus PV (kgCO$_2$eq.)	600,833,411.79	721,432,897.53	174	Propulsion units and parts for space vehicles (kgCO$_2$eq./kUSD)
GHG emissions of modules with PV (kgCO$_2$eq.)	603,931,121.79	726,562,897.53		
R&D GHG emissions (kgCO$_2$eq.)	604,286,477.38	518,270,759.72	250	Research, development, and testing services (kgCO$_2$eq./kEU)
Mission Program Management, Systems Engineering, and Mission Assurance GHG emissions (kgCO$_2$eq.)	380,179,931.11	422,675,718.09	163.11	Marketing, research and other miscellaneous professional, scientific, and technical support services (kgCO$_2$eq./kUSD)

64

Key input parameters	RD1	RD2	kgCO₂eq.	Source
Total Develop GHG emissions (kgCO₂eq.)	1,588,397,530.28	1,667,509,375.34		

Key input parameters	RD1	RD2	kgCO₂eq.	EIO-LCA Name
		Assemble		
Total Prop Needed for Starship (kg)	819,539.3301	819,539.3301	1290	/kUSD, Basic organic chemicals
Prop cost for Starship ($/kg)	$0.20	$0.20		
Single Starship Launch Emissions	3,189,600.00	3,189,600.00	MT 2683 CO_2 + 1.7 NO (*298 for CO₂eq.)	
Launcher metal mass (kg)	61,200.00	61,200.00	1.7	Steel, unspecified process
Structure composite mass (kg)	6,800.00	6,800.00	495.06	Composite-based manufactured products
Harness mass (kg)	5,950.00	5,950.00	5.84	Copper unspecified process
Electronics mass (kg)	3,400.00	3,400.00	359.46	/kUSD, Other electronic equipment
Electronics Structure mass (kg)	7,650.00	7,650.00		
Electronics area (m2)	3,929.20	3,929.20	606.12	Printed circuit board
Assemble Starship launches	767.00	1,313.00		
Assemble Starships manufactured	66.00	113.00		
Total Prop needed Assemble launches (kg)	628,586,666.21	1,076,055,140.46		
Total Prop cost Assemble launches ($)	$125,717,333.24	$215,211,028.09		

Key input parameters	RD1	RD2	kgCO$_2$eq.	EIO-LCA Name
Total Assemble launcher metal mass (kg)	4,039,200.00	6,915,600.00		
Total Assemble structure composite mass (kg)	448,800.00	768,400.00		
Total Assemble launcher harness mass (kg)	392,700.00	672,350.00		
Total Assemble launcher electronics mass (kg)	224,400.00	384,200.00		
Total Assemble launcher Electronics Structure mass (kg)	504,900.00	864,450.00		
Total Assemble launcher Electronics area (m2)	259,327.03	443,999.31		
Total Assemble launch impact (kgCO$_2$eq.)	2,997,124,796.83	5,130,771,037.99		
Assemble Mission Operations and Data Analysis (kgCO$_2$eq.)	17,420,148.00	29,751,264.00	163.11	Marketing, research and other miscellaneous professional, scientific, and technical support services (kgCO$_2$eq./kUSD)
Quantity of Servicers – Assemble	17.00	15.00		
Servicer metal mass (kg)	199.48	199.48		
Servicer structure composite mass (kg)	85.49	85.49		
Servicer harness mass (kg)	142.49	142.49		
Servicer electronics mass (kg)	698.18	698.18		
Servicer electronics Structure mass (kg)	299.22	299.22		
Total Assemble servicer metal mass (kg)	3,391.17	2,992.21		

66

Key input parameters	RD1	RD2	kgCO$_2$eq.	EIO-LCA Name
Total Assemble servicer structure composite mass (kg)	1,453.36	1,282.37		
Total Assemble servicer harness mass (kg)	2,422.26	2,137.29		
Total Assemble servicer electronics mass (kg)	11,869.09	10,472.72		
Total Assemble electronics Structure mass (kg)	5,086.75	4,488.31		
Total Assemble servicer electronics area (m2)	6,029.22	5,319.90		
Total Assemble servicer impact (kgCO$_2$eq.)	4,393,838.42	3,876,916.25		
Mission Program Management, Systems Engineering, and Mission Assurance GHG emissions (kgCO$_2$eq.)	444,381,007.87	402,109,216.48	163.11	Marketing, research and other miscellaneous professional, scientific, and technical support services (kgCO$_2$eq./kUSD)
Total Assemble GHG emissions (kgCO$_2$eq.)	3,463,319,791.13	5,566,508,434.72		

Key input parameters	RD1	RD2	kgCO$_2$eq.	EIOLCA Name
		Operate		
Ground segment cost ($)	$3,119,000,000.00	$8,270,000,000.00	182.89	/kUSD, Wireless communication equipment
Rectenna cost ($)	$623,800,000.00	$1,654,000,000.00		
Percent of mass added to support rectenna (%)	0.12244898	0.051724138		
Total rectenna support structure mass (kg)	381918367.3	427758620.7		Assumed steel

Key input parameters	RD1	RD2	kgCO$_2$eq.	EIOLCA Name
Concrete Building surface area (m2)	800	4,000	825	/m2 Industrial building, concrete and assuming similar sized buildings
Total ground physical infrastructure impact (kgCO$_2$eq.)	764,008,006.49	1,032,989,715.17		
Operate Mission Operations and Data Analysis (kgCO$_2$eq.)	70,463,520.00	70,463,520.00	163.11	Marketing, research and other miscellaneous professional, scientific, and technical support services (kgCO$_2$eq./kUSD)
Mission Program Management, Systems Engineering, and Mission Assurance GHG emissions (kgCO$_2$eq.)	8,033,167.50	21,734,407.50	163.11	Marketing, research and other miscellaneous professional, scientific, and technical support services (kgCO$_2$eq./kUSD)
Spacecraft shielding (kgCO$_2$eq.)	162,393.00	277,500.00	3.75	Glass fibers, high strength, unspecified process (kgCO$_2$eq./kg)
Total Operate GHG emissions (kgCO$_2$eq.)	842,667,086.99	1,125,465,142.67		

Key input parameters	RD1	RD2	kgCO$_2$eq.	EIOLCA Name
		Maintain		
Maintain Launches	1,534	2,613		
Maintain Starships Built	131	224		
Total Prop needed Maintain launches (kg)	1,257,173,332.42	2,141,456,269.62		
Total Prop cost Maintain launches ($)	$251,434,666.48	$428,291,253.92		
Total Maintain launcher metal mass (kg)	8,017,200.00	13,708,800.00		

68

Key input parameters	RD1	RD2	kgCO$_2$eq.	EIOLCA Name
Total Maintain structure composite mass (kg)	890,800.00	1,523,200.00		
Total Maintain launcher harness mass (kg)	779,450.00	1,332,800.00		
Total Maintain launcher electronics mass (kg)	445,400.00	761,600.00		
Total Maintain launcher Electronics Structure mass (kg)	1,002,150.00	1,713,600.00		
Total Maintain launcher Electronics area (m2)	514,724.87	880,140.23		
Total Maintain launch impact (kgCO$_2$eq.)	5,988,362,832.50	10,205,555,018.74		
Total Maintain Servicers	34	30		
Total Maintain servicer metal mass (kg)	6,782.34	5,984.41		
Total Maintain servicer structure composite mass (kg)	2,906.72	2,564.75		
Total Maintain servicer harness mass (kg)	4,844.53	4,274.58		
Total Maintain servicer electronics mass (kg)	23,738.17	20,945.45		
Total Maintain electronics Structure mass (kg)	10,173.50	8,976.62		
Total Maintain servicer electronics area (m2)	12,058.43	10,639.79		
Total Maintain servicer impact (kgCO$_2$eq.)	8,787,676.84	7,753,832.51		
Total Maintain PV area (m2)	22,946,000.00	38,000,000.00		

69

Key input parameters	RD1	RD2	kgCO₂eq.	EIOLCA Name
Maintain PV GHG emissions (kgCO₂eq.)	6,195,420.00	10,260,000.00		
Cost of Maintain modules minus PV	$3,113,493,938.78	$3,738,329,336.52		
GHG emissions of Maintain modules minus PV (kgCO₂eq.)	541,747,945.35	650,469,304.56		
GHG emissions of Maintain modules with PV (kgCO₂eq.)	547,943,365.35	660,729,304.56		
Maintain Mission Operations and Data Analysis (kgCO₂eq.)	34,840,296.00	59,502,528.00		
Mission Program Management, Systems Engineering, and Mission Assurance GHG emissions (kgCO₂eq.)	866,121,037.33	860,859,041.04	163.11	Marketing, research and other miscellaneous professional, scientific, and technical support services (kgCO₂eq./kUSD)
Total Maintain GHG emissions (kgCO₂eq.)	7,446,055,208.03	11,794,399,724.84		

Key input parameters	RD1	RD2
Dispose		
ADR Vehicles	59.00	100.00
Total Dispose servicer metal mass (kg)	11,769.35	19,948.04
Total Dispose servicer structure composite mass (kg)	5,044.01	8,549.16
Total Dispose servicer harness mass (kg)	8,406.68	14,248.60
Total Dispose servicer electronics mass (kg)	41,192.71	69,818.15
Total Dispose electronics Structure mass (kg)	17,654.02	29,922.07
Total Dispose servicer electronics area (m2)	20,924.93	35,465.98

Key input parameters	RD1	RD2
ADR vehicles GHG emissions	15,249,203.93	25,846,108.35
ADR Launches	20.00	34.00
ADR Starships Built	2	3
Total Prop needed Dispose launches (kg)	16,390,786.60	27,864,337.22
Total Prop cost Dispose launches ($)	$3,278,157.32	$5,572,867.44
Total Dispose launcher metal mass (kg)	122,400.00	183,600.00
Total Dispose structure composite mass (kg)	13,600.00	20,400.00
Total Dispose launcher harness mass (kg)	11,900.00	17,850.00
Total Dispose launcher electronics mass (kg)	6,800.00	10,200.00
Total Dispose launcher Electronics Structure mass (kg)	15,300.00	22,950.00
Total Dispose launcher Electronics area (m2)	7,858.39	11,787.59
Total Dispose launch impact (kgCO$_2$eq.)	79,794,345.28	133,295,682.50
ADR Operations GHG emissions (kgCO$_2$eq.)	763,952,825.32	1,194,230,666.73
Total Dispose GHG emissions (kgCO$_2$eq.)	828,497,966.66	1,301,680,240.88

Output	RD1	RD2
System Total		
Total GHG emissions (kgCO$_2$eq.)	14,168,937,583.09	21,455,562,918.45
GHG emissions per kWh (gCO$_2$eq./kWh)	26.58	40.38

71

Climate Comparisons

We compare SBSP systems to other baseload renewable electricity production technologies, as well as nuclear power. Wind power without storage is included for comparison as the lowest cost and emissions intensive technology tracked by NREL. GHG emissions comparisons are drawn from NREL's Lifecycle Assessment Harmonization effort (National Renewable Energy Laboratory, 2023).

Table 25. GHG emissions Comparison. Source data from NREL: surveying thousands of publications, NREL documents the range of assessed GHG emissions, denoting low, median, and high impact assessment results for each technology. All units are in $gCO_2eq./kWh$.

Technology ($gCO_2eq./kWh$)	Low Impact	Median Impact	High Impact
RD1 Baseline		26.58	
RD2 Baseline		40.38	
Nuclear	5	13	210
Hydropower	4	21	90
Utility-Scale Solar Photovoltaics	25	43	190
Concentrating Solar Power	10	28	90
Onshore Wind	6	13	50
Offshore Wind	6	13	50
Tidal	5	8	12
Geothermal	4	37	230
Natural Gas	350	486	950
Coal	840	1001	1690

Sensitivity Analyses

We performed several sensitivity analyses to assess the necessary conditions for commercially viable SBSP systems as compared to 2050 projections for existing renewable electricity production technologies. Parameters that could reasonably be expected to vary with uncertainty were considered; these relate to launch, manufacturing, and solar cell efficiency. After varying individual variables separately, we conclude with a multiple-variable sensitivity analysis which considers

72

impacts of all manufacturing, launch, and solar cell modifications. To recreate these results, simply take the above data and relationships in calculation and modify as described below.

Launch

Sensitivity Analysis 1 - Reduced Launch Costs 1 a) $50M b) $10M

- Summary of modification(s): Reduction of Starship launch costs from $100,000,000 per launch to a) $50,000,000 and b) $10,000,000 per launch.
- Rationale: Baseline launch costs per kg take the lowest recorded cost (Falcon Heavy, 2019, $1500), and apply the same decrease in launch prices from the previous 10 years. SpaceX has announced that Starship launches might reduce to $10,000,000 per launch in the 2024-2025 timeframe[i]. Additionally, launch costs declined 36% in the past 10 years. Based on these expectations and assuming launches in 2040, we consider medium and low scenarios for launch, which is the main driver of both SBSP system costs (Duffy, 2022).
- Result: For $50M, this results in a 32% total cost savings for the Innovative Heliostat Swarm and a 36% total cost savings for the Planar Array. For $10M, this results in a 61% reduction in cost for Innovative Heliostat Swarm, and 69% for Mature Planar Array.

Sensitivity Analysis 2 - Launch Direct to GEO

- Summary of modification(s): Starships take payloads directly to GEO, removing refuel launches and reusability, and reducing payload capacity to 21 MT.
- Rationale: Starship is able to deliver a fraction of its payload to GEO; this is a test of the costs and benefits of this approach.
- Result: While costs are reduced by at least 42% and 47% for RD1 and RD2, respectively, due to the reduced number of launches, GHG emissions decrease just 8% and 9%, respectively, due to the increased number of Starships manufactured.

Sensitivity Analysis 3 - Electric Propulsion for Orbital Transfer

- Summary of modification(s): We added 17.2% of system mass and manufacturing cost based on propulsion mass per module, and removed launches to refuel in LEO. This modification is based on first order delta v calculations using specifications from NASA's NEXT-C thruster (NASA, 2023), assuming >4,700m/s is needed to reach GEO from LEO.
- Rationale: Though using EP adds upmass, it can significantly reduce the number of launches required, as there is no need to refuel in LEO. This makes the most use out of Starship's payload capacity to LEO, but comes at a cost of a longer time (about 4 months) to complete the orbital transfer.

- Result: This results in a 62% total cost savings for the Innovative Heliostat Swarm and a 69% total cost savings for the Mature Planar Array. The travel time from LEO to GEO increased to four months as opposed to refueling for one month and traveling for one month under chemical propulsion. The additional mass of the propulsion units did not appreciably change assembly time.

Manufacturing

Sensitivity Analysis 1 - Initial Hardware Costs

- Summary of modification(s): We apply a 10% reduction in initial module costs, first servicer costs, and ADR vehicle base cost. The reduction in hardware costs affects all other parameters that use these costs as inputs, such as program management and technology development.
- Rationale: Original hardware costs for this analysis were based on U.S. Government technology demonstration missions. We reduce hardware development costs to levels provided by a survey of existing and upcoming commercial offerings performed by Aerospace.
- Result: Combining all reduced costs results in an 19% total cost savings to the Innovative Heliostat Swarm and 14% total cost savings to the Mature Planar Array.

Sensitivity Analysis 2 - Learning Curve Reduction

- Summary of modification(s): learning rate for servicers reduced from 85% to 80%, learning rate for modules reduced from 75% to 70%, and learning rate for ADR vehicles reduced from 90% to 80%.
- Rationale: Aerospace consistently used a learning curve model based on cumulative averages, which results in faster learning when compared to the model used for this analysis. This method leverages data from other industries with complex processes and production of 100s to 1,000s of units to evaluate learning rate and cost combinations. We therefore decrease our learning rates to be on par with this method while still considering the units-based Crawford model.
- Result: Considering all updated learning rates results in a 4% total cost savings for the Innovative Heliostat Swarm and a 3% total cost savings for the Mature Planar Array.

Sensitivity Analysis 3 - Increased Hardware Lifetime

- Summary of modification(s): Baseline module and servicer lifetime is set to 10 years. Here, the lifetime is increased to 15 years.
- Rationale: Today's GEO hardware is most commonly baselined for a 15-year lifetime.
- Result: The extended lifetime leads to a reduction of one refurbishment cycle, saving 21% in fiscal costs for Innovative Heliostat Swarm and 23% in GHG emissions per kWh. For Mature Planar Array, the cost reduction was 29% and emissions reduction was 28%.

Solar Cell Efficiency

Sensitivity Analysis 1 - Solar Cell Efficiency (15% Increase)

- Summary of modification(s): Efficiency of solar cells is increased by 15 percentage points, from 35% to 50%.
- Rationale: NASA noted that the state of the practice for solar cell efficiency is 33%, while the state of the art is 70% (NASA, 2022). The best research cell efficiency currently tracked by NREL is 47.6% (NREL, n.d.).
- Result: The improvement in solar cell efficiency leads to a reduction in mass to deliver the same 2 GW of power, and therefore less manufacturing and fewer launches, resulting in 19% cost savings for Innovative Heliostat Swarm, 23% for Mature Planar Array, and 25% GHG emissions reduction per kWh for Innovative Heliostat Swarm and 28% for Mature Planar Array.

Multiple Variable Sensitivity Analyses

Sensitivity Analysis 1 – Competitive Solution

- Summary of modification(s):
  - Launch costs are here reduced from $100M to $50M.
  - Efficiency of solar cells is increased by 15%.
  - Initial servicer and ADR vehicle costs are reduced by 90%.
  - Learning curves are lowered by 5 percentage points across the board.
  - 1,720kg of EP (fuel + hardware) for every 10,000kg of system mass for orbital transfer; 17.2% additional cost to total module manufacturing cost.
- Rationale: Select sensitivities were combined with modifications to understand what conditions may lead the SBSP design references to become economically and environmentally competitive with existing renewable electricity technology projections for 2050.
- Result: Costs are reduced by 95% for both systems, and emissions intensity by 86% for Innovative Heliostat Swarm and 90% for Mature Planar Array.

Sensitivity Analysis 2 - Present Day Costs Only

- Summary of modification(s):
  - Launch costs are here increased from $100M to $150M.
  - Initial module cost is increased to $5M
  - Learning curves are increased to 85% for modules and 95% for servicers and ADR vehicles.
  - Operations costs are increased from a flat fee to $500k per 2,800kg of system mass annually.
  - Starship reuses are reduced from 100 to 50.

75

- Rationale: It is generally understood that SBSP is highly cost prohibitive with today's costs, and technically infeasible with today's technology. Simply for comparison, we apply more of today's costs and capabilities.
- Result: Fiscal costs are increased 5.3 times for RD1 and 4.5 times for RD2. Emissions per kWh for both systems increase by a factor of 10.

Table 26. All Sensitivity Analysis Results

Sensitivity Analysis Grouping	Technology / Scenario	LCOE ($/kWh)	Emissions Intensity (gCO$_2$eq./kWh)
Baseline	RD1 Baseline	0.6	26.6
	RD2 Baseline	1.6	40.4
Launch	RD1 Lower Launch Costs ($50M)	0.4	26.6
	RD1 Lower Launch Costs ($10M)	0.2	26.6
	RD2 Lower Launch Costs ($50M)	1	40.4
	RD2 Lower Launch Costs ($100M)	0.5	40.4
	RD1 Direct to GEO	0.3	24.4
	RD2 Direct to GEO	0.8	36.6
	RD1 Electric Propulsion Orbital Transfer	0.2	11.4
	RD2 Electric Propulsion Orbital Transfer	0.5	14.2
Manufacturing	RD1 Extended Component Lifetime	0.5	19.3
	RD2 Extended Component Lifetime	1.2	28.8
	RD1 Lower Hardware Cost	0.5	19.5
	RD2 Lower Hardware Cost	1.3	32.3
	RD1 Lower Manufacturing Learning	0.6	22.6

Sensitivity Analysis Grouping	Technology / Scenario	LCOE ($/kWh)	Emissions Intensity (gCO$_2$eq./kWh)
	RD2 Lower Manufacturing Learning	0.8	35.8
Solar Cell Efficiency	RD1 Efficient Solar Cells	0.5	20.4
	RD2 Efficient Solar Cells	1.2	29.6
	RD1 'Competitive'	0.03	3.6
Multiple Variables	RD2 'Competitive'	0.08	4.2
	RD1 'Present Day'	4.18	286
	RD2 'Present Day'	10.73	360

77

Appendix C: Acknowledgements

This appendix lists individuals and organizations who helped inform this study's discussion of SBSP, whether through interviews, source material, reviews, or informal discussions. The authors of this study are extremely grateful for the time provided by the leading experts and institutions willing to impart their humbling level of expertise in the performance of this exploratory study.

John Scott (NASA STMD)

Heath Garvin (NREL)

Roger Myers

Paul Jaffe (NRL/USDR&E)

James Winter (AFRL)

John Mankins (Artemis Innovation Management Solutions)

Wes Furnham (Applied Physics Laboratory)

Murat Okandan (mPower)

Alfred Nash (NASA JPL)

Steve Matousek (NASA JPL)

Matthew Weinzierl and team (Harvard)

John Bucknell and team (Virtus Solis)

Sergio Pellegrino and team (CalTech)

Brent Sherwood (Blue Origin)

Peter Garretson (American Foreign Policy Council)

Olga Spahn (Advanced Research Projects Agency-Energy, ARPA-E)

Guilherme Larangueira (ARPA-E)

Igor Cvetkovic (ARPA-E)

Sanjay Vijendran (ESA)

Leopold Summerer (ESA)

Advenit Makaya (ESA)

Zhiyuan Fan and team member (Columbia Center on Global Energy Policy)

Douglas Hollett (DoE)

Rima Oueid (DoE)

Alex Ellery (Carleton University)

Milo McBride (New York University Tandon School of Engineering)

Gavin Schmidt (NASA Goddard Institute for Space Sciences)

Paul Stackhouse (NASA Science Mission Directorate)

The information contained in this report reflects the views and opinions of the authors and do not necessarily represent the views or opinions of, nor should be construed as endorsements by the individuals or organizations listed in Appendix C.

Appendix D: Acronyms & Key Terms

Acronym or Term	Definition
ADR	Active debris removal
AFRL	Air Force Research Laboratory
ATB	Annual Technology Baseline
Capacity factor (%)	Maximum fractional portion of the year the system is producing power
CapEx	Capital Expenditures
CapEx ($/kW)	Initial capital expenditures to produce the system, per unit of energy generation
CO_2	Carbon Dioxide
CO_2	Carbon Dioxide
ConOps	Concept of Operations
DARPA	Defense Advanced Research Projects Agency
DC	Direct Current
DoD	Department of Defense
DoE	Department of Energy
DSA	Distributed Spacecraft Autonomy

Acronym or Term	Definition
EIA	Energy Information Agency
EIO-LCA	Economic Input Output – Lifecycle Analysis
EP	Electric Propulsion
EPA	Environmental Protection Agency
ESA	European Space Agency
EU	European Union
Fixed charging rate (%)	Annualized cost of a capital for the system (similar to a "discount rate")
FOAK	First-of-a-kind
FOM	Fixed operations and maintenance
FOM ($/kw-year)	Fixed annual expenditures for operations and maintenance per unit of generation
gCO$_2$eq.	Grams of carbon dioxide equivalent per kilowatt-hour
GEO	Geostationary Orbit
GHG	Greenhouse gas
GHz	Gigahertz
GPIM	Green Propellant Infusion Mission

Johnson, E. M., & Roulette, J. (2018, October 31). *Musk Shakes Up SpaceX In Race To Make Satellite Launch Window: Sources*. Retrieved from Reuters: https://www.reuters.com/article/us-spacex-starlink-insight-idUSKCN1N50FC

Kall Morris Inc. (2023, June). *Keeping Space Clear for All*. Retrieved June 2023, from https://www.kallmorris.com/

Kall Morris Inc. (2023). *Keeping Space Clear for All*. Retrieved from https://www.kallmorris.com/

Kawahara, S. (2023, May 27). *Japan to try beaming solar power from space in mid-decade*. Retrieved from Nikkei: https://asia.nikkei.com/Business/Science/Japan-to-try-beaming-solar-power-from-space-in-mid-decade

Kirschner, K. (2023, June 21). *Orbital Composites, Virtus Solis Team on Space-Based Solar Power Station*. Retrieved from https://payloadspace.com/orbital-composites-virtus-solis-team-on-space-based-solar-power-station/

Leontif, W. W. (1951). Wassily Leontief-Input-Output Economics. *Scientific American*, 12-21. Retrieved from https://cooperative-individualism.org/leontief-wassily_input-output-economics-1951-oct.pdf

Liang, H. a. (2021). Reshoring Silicon Photovoltaics Manufacturing Contributes To Decarbonization And Climate Change Mitigation. *Nature Communications*, 1-19. doi:https://doi.org/10.1038/s41467-023-36827-z

Liang, H., & You, F. (2023, March). Retrieved from Nature: https://www.nature.com/articles/s41467-023-36827-z

Longshot Space. (n.d.). Retrieved from https://www.longshotspace.com/

Lubin, P. (2021, April 2). *Moonbeam-Beamed Lunar Power*. Retrieved from https://www.nasa.gov/directorates/spacetech/strg/lustr/2020/Moonbeam_Beamed_Lunar_Power/

Mankins, J. C. (1997). A fresh look at space solar power: New architectures, concepts and technologies. *Acta Astronautica*, 347-359.

Mankins, J. C. (1997). A fresh look at space solar power: New architectures, concepts and technologies. *Acta Astronautica*, 347-359.

Mankins, J. C. (2012). *SPS-ALPHA: The First Practical Solar Power Satellite via Arbitrarily Large Phased Array*. Retrieved from https://www.nasa.gov/pdf/716070main_Mankins_2011_PhI_SPS_Alpha.pdf

Mankins, J. C. (2021). SPS-Alpha Mark-III and an achievable roadmap to space solar power. *72nd international astronautical congress*.

87

Acronym or Term	Definition
GTO	Geostationary Transfer Orbit
IAEG	International Aerospace Environmental Group
ISAM	In-Space Assembly Manufacturing
ISS	International Space Station
ITU	International Telecommunications Union
JAXA	Japan Aerospace Exploration Agency
KARI	Korean Aerospace Research Institute
KERI	Korean Energy Research Institute
kg / Mkg	Kilogram / Million kilograms
kUSD	Thousands of U.S. dollars
kW / MW / GW	Kilowatt / Megawatt / Gigawatt
kWh / MWh	Kilowatt-hour / Megawatt-hour
LCOE	Levelized Cost of Electricity
LEO	Low Earth Orbit
Lifecycle	A series of stages through which something passes during its lifetime

Acronym or Term	Definition
Lifecycle	A series of stages through which something passes during its lifetime.
MEV	Mission Extension Vehicle
MMmt	Million Metric Tons
MOCET	Mission Operations Cost Estimation Tool
MT	Metric ton
NAICS	North American Industry Classification System
NASA	National Aeronautics and Space Administration
NIAC	NASA Innovative Advanced Concepts
NOAA	National Oceanic and Atmospheric Administration
NREL	National Renewable Energy Laboratory
NRL	Naval Research Laboratory
O&M	Operations and Maintenance
OGA	Other Government Agency
OSAM	On-Orbit Servicing, Assembly, and Manufacturing
OSTP	Office of Science and Technology Policy
OTPS	Office of Technology, Policy, and Strategy

Acronym or Term	Definition
OTV	Orbital Transfer Vehicle
PV	Photovoltaic
R&D	Research and Development
RD1, RD2	Representative Design 1, Representative Design 2
Renewables	Electricity technologies that do not consume fossil fuels (oil, coal, gas) as a primary input for production
Renewables	Electricity technologies that do not consume fossil fuels (oil, coal, gas) as a primary input for production
RF	Radio frequency
RFP	Request for proposals
RPO	Rendezvous and Proximity Operations
SBSP	Space-Based Solar Power
SME	Subject Matter Expert
SSPIDR	Space Solar Power Incremental Demonstrations and Research Project
STMD	Space Technology Mission Directorate
TRL	Technology Readiness Level
U.S.	United States

Acronym or Term	Definition
UK	United Kingdom
UN	United Nations
VOM ($/kWh)	Variable operations and maintenance are expenditure per unit of generation for operations and maintenance
WBS	Work Breakdown Structure

References

Aerospace. (2018). Learning Rate Sensitivity Model. *2018 NASA Cost and Schedule Symposium.* Retrieved from https://www.nasa.gov/sites/default/files/atoms/files/27_learning_rate_sensitivity_model-2018_nasa_cost_symposium.pdf

Blue Origin Fed'n, LLC; Dynetics, Inc.-A Leidos Co., B-419783 et al. (CPD ¶ 265 at 27 n.13, July 30, 2021).

Boeing. (2012). *702SP Satellite.* National Aeronautics and Space Administration. Retrieved from https://rsdo.gsfc.nasa.gov/images/catalog/301406_RapidIII_4Page_Layout_A_REV_1_11.pdf

Bouckaert, S., Pales, A., McGlade, C., Remme, U., Wanner, B., Varro, L., & Spencer, T. (2021). *Net Zero by 2050: A Roadmap for the Global Energy Sector.*

Brandon, E. (2019, April 10). *Power Beaming for Long Life Venus Surface Missions*. Retrieved June 2023, from https://www.nasa.gov/directorates/spacetech/niac/2019_Phase_I_Phase_II/Power_Beaming/

Brandon, E., Bugga, R., Grandidie, J., Jeff, H. L., Schwartz, J. A., & Limaye, S. (2020, March). *Power Beaming for Long Life Venus Surface Missions, FInal Report*. Retrieved from https://www.nasa.gov/directorates/spacetech/niac/2019_Phase_I_Phase_II/Power_Beaming/

Brunner, K., & Jack, J. (2006). TRL Impact on Cost. *Joint SCAF SSCAG, & EACE International Conference*, (pp. 19-21).

BryceTech. (2022). *State of the Satellite Industry.* Satellite Industry Association.

BryceTech. (2023). *2022 Orbital Launches Year in Review*. Retrieved from https://brycetech.com/reports/report-documents/Orbital_Launches_Year_in_Review_2022.pdf

Caltech. (2023, June 1). *In a First, Caltech's Space Solar Power Demonstrator Wirelessly Transmits Power in Space*. Retrieved from https://www.caltech.edu/about/news/in-a-first-caltechs-space-solar-power-demonstrator-wirelessly-transmits-power-in-space

Center for Sustainable Systems. (2022). *Photovoltaic Energy Fact Sheet*. Retrieved from https://css.umich.edu/publications/factsheets/energy/photovoltaic-energy-factsheet

Cobas-Flores, E. C. (1998). Economic Input-Output-Based Life-Cycle Assessment (EIO-LCA). Retrieved from https://www.researchgate.net/publication/242142910_Economic_Input-Output-Based_Life-Cycle_Assessment_EIO-LCA

DARPA. (2022, October 10). *POWER Aims to Create Revolutionary Power Distribution Network*. Retrieved from https://www.darpa.mil/news-events/2022-10-05b

de Selding, P. B. (2013, June 20). *Electric-propulsion Satellites Are All the Rage*. Retrieved from SpaceNews: https://spacenews.com/35894electric-propulsion-satellites-are-all-the-rage/

Duffy, K. (2022, February 11). *Elon Musk Says He's 'Highly Confident' That SpaceX's Starship Rocket Launches Will Cost Less Than $10 Million Within 2-3 Years*. Retrieved from Business Insider: https://www.businessinsider.com/elon-musk-spacex-starship-rocket-update-flight-cost-million-2022-2

EIA. (2022, March 18). *EIA projects that renewable generation will supply 44% of U.S. electricity by 2050*. Retrieved from https://www.eia.gov/todayinenergy/detail.php?id=51698

EIA. (2023, May 21). *How Much of U.S. Carbon Dioxide Emissions Are Associated With Electricity Generation?* Retrieved from https://www.eia.gov/tools/faqs/faq.php?id=77&t=11

EPA. (2023, April 28). *Sources of Greenhouse Gas Emissions*. Retrieved from https://www.epa.gov/ghgemissions/sources-greenhouse-gas-emissions

Frazer Nash Consultancy. (2022). *Space-Based Solar Power as a Contributor to Net Zero.* Retrieved from https://www.esa.int/Enabling_Support/Space_Engineering_Technology/SOLARIS/Cost_vs._benefits_studies

Frazer-Nash Consultancy; London Economics. (2022). *Space-Based Solar Power.* European Space Agency.

Gangestad, J. W. (2017). *Orbital Slots for Everyone?* Retrieved from Aerospace Corporation: https://aerospace.org/sites/default/files/2018-05/OrbitalSlots_0.pdf

GmbH, Roland Berger. (2022). *Space-Based Solar Power.* Retrieved from https://www.esa.int/Enabling_Support/Space_Engineering_Technology/SOLARIS/Cost_vs._benefits_studies

Hamisevicz, M. E. (2023, July 13). *U.S. Naval Research Laboratory*. Retrieved from First In-Space Laser Power Beaming Experiment Surpasses 100 Days of Successful On-Orbit Operations: https://www.nrl.navy.mil/Media/News/Article/3457014/first-in-space-laser-power-beaming-experiment-surpasses-100-days-of-successful/

Holt, L. C. (2022, November 4). *Air Force Research Laboratory*. Retrieved from SpaceWERX awards 124 Orbital Prime contracts: https://www.afrl.af.mil/News/Article-Display/Article/3210527/spacewerx-awards-124-orbital-prime-contracts/

International Aerospace Environmental Group. (2023, May 21). *GHG Reporting Guidance*. Retrieved from https://www.iaeg.com/ghg-guidance

Marshall, M. A., Madonna, R. G., & Pellegrino, S. (2023). Investigation of Equatorial Medium Earth Orbits for Space Solar Power. *Transaction on Aerosapce And Electronic Systems*, Pre-Publication. Retrieved from Investigation of Equatorial Medium Earth Orbits for

McFadden, C. (2023, May 1). *DARPA is making Nikola Tesla's dream of Wireless Energy a Reality*. Retrieved June 2023, from https://interestingengineering.com/innovation/darpa-laser-power-transfer

Meisl, P., & Morales, L. (1994). *Historical Cost Improvement Curves for Selected Satellites: Final Report*. Management Consulting and Research, Inc. Retrieved from TR-9338/029-1

Messier, D. (2020, June 2). *Over Budget Restore-L Mission 3.5 Years from Launch*. Retrieved from Parabolic Arc.

NASA & DoE. (1980). Final Proceedings Solar Power Satellite Program Review. Retrieved from https://ntrs.nasa.gov/api/citations/19820014802/downloads/19820014802.pdf

NASA. (2020, October 23). *Distributed Spacecraft Autonomy (DSA)*. Retrieved from https://www.nasa.gov/directorates/spacetech/game_changing_development/projects/dsa

NASA. (2020, December). *Lightweight Materials and Structures*. Retrieved from https://www.nasa.gov/directorates/spacetech/game_changing_development/projects/archived/LMS

NASA. (2022, June 23). *On-Orbit Servicing, Assembly, and Manufacturing 2*. Retrieved from https://www.nasa.gov/mission_pages/tdm/osam-2.html

NASA. (2022). *State of the Art of Small Spacecraft Technology, 3.0 Power*. Retrieved from https://www.nasa.gov/smallsat-institute/sst-soa/power

NASA. (2022). *Superlightweight Aerospace Composites*. Retrieved from https://www.nasa.gov/directorates/spacetech/game_changing_development/projects/sac

NASA. (2023, July). *2023 Tipping Point Selections*. Retrieved from https://www.nasa.gov/directorates/spacetech/solicitations/tipping_points/2023_selections

NASA. (2023). *Earth Orbit Environmental Heating. Lesson 693*. Retrieved from https://llis.nasa.gov/lesson/693.

NASA. (2023, January). *Gridded Ion Thrusters (NEXT-C)*. Retrieved from https://www1.grc.nasa.gov/space/sep/gridded-ion-thrusters-next-c/

NASA Library. (2023, May 19). *Space-Based Solar Power*. Retrieved from Web of Science: https://www.webofscience.com/wos

National Aeronautics and Space Administration. (2006). *NASA Correspondence Management and Communications Standards and Style.*

National Aeronautics and Space Administration. (2015, June 10). *Green Propellants*. Retrieved June 2023, from https://www.nasa.gov/centers/wstf/testing_and_analysis/propellants_and_aerospace_fulids/green_propellants.html

National Renewable Energy Laboratory. (2023). *Life Cycle Assessment Harmonization*. Retrieved May 2023, from https://www.nrel.gov/analysis/life-cycle-assessment.html

NOAA. (2022, June 22). *Projected increase in space travel may damage ozone layer*. Retrieved from NOAA Research: https://research.noaa.gov/2022/06/21/projected-increase-in-space-travel-may-damage-ozone-layer/

Northrop Grumman. (2023, June). *SpaceLogistics*. Retrieved June 2023, from https://www.northropgrumman.com/space/space-logistics-services/

Northrop Grumman. (2023). *SpaceLogistics*. Retrieved from https://www.northropgrumman.com/space/space-logistics-services/.

NREL. (2022). *Annual Technology Baseline*. Retrieved from https://atb.nrel.gov/

NREL. (2022). *Definitions*. Retrieved from https://atb.nrel.gov/electricity/2022/definitions#capex.

NREL. (2023). *Best Research-Cell Efficiency Chart*. Retrieved from https://www.nrel.gov/pv/cell-efficiency.html

NREL. (2023). *Simple Levelized Cost of Energy (LCOE) Calculator Documentation*. Retrieved from https://www.nrel.gov/analysis/tech-lcoe-documentation.html.

NREL. (n.d.). *Best Research Cell Efficiency Chart*. Retrieved from https://www.nrel.gov/pv/cell-efficiency.html

OneWeb. (2023). *OneWeb*. Retrieved from https://oneweb.net/

OneWeb. (2023, June). *OneWeb*. Retrieved June 2023, from https://oneweb.net/

Orbit Fab. (2023, June). *Refuel Your Spacecraft*. Retrieved June 2023, from https://www.orbitfab.com/

OrbitFab. (2023). *Refuel your Spacecraft*. Retrieved from https://www.orbitfab.com/

P.R. Shukla, J. S. (2022). *IPCC 2022, Summary for Policymakers.* Cambridge, UK and New York, NY, USA: Cambridge University Press. doi:10.1017/9781009157926.001

Pellegrino. (2022). A lightweight space-based solar power generation and transmission satellite. Retrieved from https://doi.org/10.48550/arXiv.2206.08373

Radiative efficiency of state-of-the-art photovoltaic cells. (2012, June). *Progress in Photovoltaics*, 472-476. Retrieved from https://onlinelibrary.wiley.com/doi/10.1002/pip.1147

Rodenbeck, C. T., Jaffe, P. I., Strassner II, B. H., Hausgen, P. E., McSpadden, J. O., Kazemi, H., . . . Self, A. P. (2020). Microwave and Millimeter Wave Power Beaming. *IEEE Journal of Microwaves*, 229-259. Retrieved from https://www.researchgate.net/publication/348442155_Microwave_and_Millimeter_Wave_Power_Beaming

Roland Berger GmbH; OHB System AG. (2022). *Space-Based Solar Power.* European Space Agency.

Sasaki, S. (2006). A new concept of solar power satellite: Tethered-SPS. *Acta Astronautics*, 153-165.

Sasaki, S. (2009). Demonstration Experiment for Tethered-Solar Power Satellite. *JSASS Space Tech Japan Vol 7*, pp. Tr_1_1-Tr_1_4.

Sheetz, M. (2020, August 10). *SpaceX is manufacturing 120 Starlink internet satellites per month*. Retrieved from Consumer News and Business Channel: https://www.cnbc.com/2020/08/10/spacex-starlink-satellite-production-now-120-per-month.html

Sheetz, M. (2020, August 10). *SpaceX is Manufacturing 120 Starlink Internet Satellites Per Month*. Retrieved June 2023, from CNBC: https://www.cnbc.com/2020/08/10/spacex-starlink-satellite-production-now-120-per-month.html

Silver Circuits. (n.d.). *Frequently Asked Questions*. Retrieved from Custom PCB: https://custompcb.com/faq.php

SpaceX. (2023, June). *SpaceX*. Retrieved May 2023, from https://www.spacex.com/

SpaceX. (2023). *Starlink*. Retrieved from https://www.starlink.com

SpinLaunch. (n.d.). Retrieved from https://www.spinlaunch.com/#p1

Starlink. (2023). *Starlink*. Retrieved May 2023, from https://www.starlink.com

United Nations. (2023, March 27). *Climate Action*. Retrieved from https://www.un.org/en/climatechange/net-zero-coalition.

United States Department of State and the United States Executive Office of the President. (2021). *The Long-Term Strategy of the United States: Pathways to Net-Zero Greenhouse Gas Emissions by 2050.* Washington, DC: United States Department of State and the United States Executive Office of the President. Retrieved from https://www.whitehouse.gov/wp-content/uploads/2021/10/US-Long-Term-Strategy.pdf

United States National Science and Technology Council. (2022). *In-Space Servicing, Assembly, Manudacturing National Strategy.* Washington DC. Retrieved from https://www.whitehouse.gov/wp-content/uploads/2022/04/04-2022-ISAM-National-Strategy-Final.pdf

Vorbach, I. (2022, May 19). *Is Starship Really Going To Revolutionize Launch Costs?* Retrieved from SpaceDotBiz: https://newsletter.spacedotbiz.com/p/starship-really-going-revolutionize-launch-costs

Web of Science. (2023). *Web of Science*. Retrieved May 2023, from https://www.webofscience.com/wos

Werner, D. (2018, August 8). *Electric propulsion to send smallsats from LEO to GEO orbit, moon.* Retrieved from https://spacenews.com/electric-propulsion-to-send-smallsats-from-leo-to-geo-orbit-moon/

_____